High Energy Particle Physics Workshop 2016

Journal of Physics: Conference Series Volume 802

Johannesburg, South Africa
8 – 10 February 2016

Editors:

Alan S. Cornell
Bruce Mellado

ISBN: 978-1-5108-3694-5
ISSN: 1742-6588

Printed from e-media with permission by:

Curran Associates, Inc.
57 Morehouse Lane
Red Hook, NY 12571

Some format issues inherent in the e-media version may also appear in this print version.

Printed by Curran Associates, Inc. (2017)

For permission requests, please contact the Institute of Physics
at the address below.

Institute of Physics
Dirac House, Temple Back
Bristol BS1 6BE UK

Phone: 44 1 17 929 7481
Fax: 44 1 17 920 0979

techtracking@iop.org

Additional copies of this publication are available from:

Curran Associates, Inc.
57 Morehouse Lane
Red Hook, NY 12571 USA
Phone: 845-758-0400
Fax: 845-758-2633
Email: curran@proceedings.com
Web: www.proceedings.com

Table of contents

Volume 802

High Energy Particle Physics Workshop 2016
8–10 February 2016, Johannesburg, South Africa

Accepted papers received: 11 January 2017
Published online: 23 February 2017

Journal of Physics: Conference Series **802** (2017) 011001 doi:10.1088/1742-6596/802/1/011001

Preface

The motivation for this workshop began with the discovery of the Higgs boson four years ago, and the realisation that many problems remain in particle physics, such as why there is more matter than anti-matter, better determining the still poorly measured parameters of the strong force, explaining possible sources for dark matter, naturalness etc. While the newly discovered Higgs boson seems to be compatible with the Standard Model, current experimental accuracy is far from providing a definitive statement with regards to the nature of this new particle. There is a lot of room for physics beyond the Standard Model to emerge in the exploration of the Higgs boson. Recent measurements in high-energy heavy ion collisions at the LHC have shed light on the complex dynamics that govern high-density quark-gluon interactions. An array of results from the ALICE collaboration has been highlighted in a recent issue of CERN courier. The physics program of high-energy heavy ion collisions promises to further unveil the intricacies of high-density quark-gluon plasma physics.

The great topicality of high energy physics research has also seen a rapid increase in the number of researchers in South Africa pursuing such studies, both experimentally through the ATLAS and ALICE colliders at CERN, and theoretically. Young researchers and graduate students largely populate these research groups, with little experience in presenting their work, and few support structures (to their knowledge) to share experiences with. Whilst many schools and workshops have sought to educate these students on the theories and tools they will need to pursue their research, few have provided them with a platform to present their work. As such, this workshop discussed the various projects being pursued by graduate students and young researchers in South Africa, enabling them to develop networks for future collaboration and discussion.

The workshop took place at the iThemba Laboratories - North facility, in Gauteng, from the 8th to the 10th of February 2016, where excellent conference facilities with outdoors and indoor tea areas for discussions and interactions were provided, along with a state-of-the-art remote access to the conference venue such that those who were unable to attend the workshop in person could also be present. The laboratory is located next door to the Wits Professional Development Hub (on the corner of Jan Smuts Avenue and Empire Road), which provided the catering for this workshop. A morning plenary session, followed 15+10 minute presentations, was the format across our three days. The topics covered being in high-energy theory and phenomenology (heavy ions, pp, ep, ee collisions), ATLAS physics and ALICE physics. The workshop website is http://hep.wits.ac.za/HEPPW2016.php

Alan S. Cornell and Bruce Mellado

Journal of Physics: Conference Series **802** (2017) 011001 doi:10.1088/1742-6596/802/1/011001

Figure 1. *The opening address delivered by Deputy Vice-Chancellor for research, Zeblon Vilakazi.*

Figure 2. *An opening address delivered by the head of the School of Physics, Joao Rodrigues.*

Figure 3. *An opening address delivered by the President of the SAIP, Azwinndini Muronga.*

Figure 4. *Introductory talk given by the head of SA-CERN, Jean Cleymans.*

Figure 5. *Welcome lecture given by Bruce Mellado.*

Figure 6. *Appreciation gift presented to John Carter.*

Sponsors

- National Institute for Theoretical Physics (NITheP)
- The University of the Witwatersrand
- iThemba Laboratories
- Mandelstam Institute for Theoretical Physics

Organising Committee

Bruce Mellado	University of the Witwatersrand - (*co-chair*)
Alan S Cornell	NITheP; University of the Witwatersrand - (*co-chair*)
Zinhle Buthelezi	iThemba Laboratories
Jean Cleymans	University of Cape Town
Tom Dietel	University of Cape Town
Dino Giovannoni	Rhodes
Betty Kibirige	University of Zululand
Will Horowitz	University of Cape Town
Deepak Kar	University of the Witwatersrand
Azwinndini Muronga	Nelson Mandela Metropolitan University
Xifeng Ruan	University of the Witwatersrand
Elias Sideras-Haddad	Materials Physics Research Institute; University of the Witwatersrand
Sahal Yacoob	University of Cape Town

List of attendees

John Carter	University of the Witwatersrand
Jean Cleymans	University of Cape Town
Alan Cornell	University of the Witwatersrand
Tom Dietel	UCT/iThemba
Nomvelo Dindikazi	UCT/iThemba
Megan Lee Ellor	University Zululand
Eilam Gross	Weizmann Institute of Science
Dino Giovannoni	Rhodes University
Gerhard Harmsen	University of the Witwatersrand
Will Horowitz	University of Cape Town
Abdullah Khalil Hassan Ibrahim	University of Cape Town

Harshna Jivan	University of the Witwatersrand
Deepak Kar	University of the Witwatersrand
Mohammed Omer Khojali	University of the Witwatersrand
Betty Kibirige	University Zululand
Isobel Kolbe	University of Cape Town
Mukesh Kumar	University of the Witwatersrand
Chamunorwa Oscar Kureba	University of the Witwatersrand
Massimiliano Marchisone	UCT/iThemba
Sizwe Jackson Clement Masuku	University Zululand
Joyful Mdhluli	University of the Witwatersrand
Benjamin Meiring	University of Cape Town
Bruce Mellado	University of the Witwatersrand
Sibaliso Mhlanga	UCT/iThemba
Tshidiso Molupe	University of the Witwatersrand
Chane Simone Moodley	University of the Witwatersrand
Razieh Morad	University of Cape Town
Chuene Mosomane	University of the Witwatersrand
Skhathisomusa Dumani Mthembu	University of the Witwatersrand
Mxolisi David Mtshali	University Zululand
Azwinndini Muronga	Nelson Mandela Metropolitan University
Chad Pelwan	University of the Witwatersrand
Andriniaina Rasoanaivo	University of Cape Town
Robert G Reed	University of the Witwatersrand
Xifeng Ruan	University of the Witwatersrand
Marc Sacks	University of the Witwatersrand
Kgotlaesele Johnson Senosi	UCT/iThemba
Matthew Spoor	University of the Witwatersrand
Kehinde Tomiwa	University of the Witwatersrand
Zeblon Vilakazi	University of the Witwatersrand
Stefan von Buddenbrock	University of the Witwatersrand
Andile Whitehead	University of Cape Town

Journal of Physics: Conference Series **755** (2016) 011001 doi:10.1088/1742-6596/755/1/011001

Peer review statement

All papers published in this volume of *Journal of Physics: Conference Series* have been peer reviewed through processes administered by the proceedings Editors. Reviews were conducted by expert referees to the professional and scientific standards expected of a proceedings journal published by IOP Publishing.

IOP Conf. Series: Journal of Physics: Conf. Series **802** (2017) 012001 doi:10.1088/1742-6596/802/1/012001

A heavy scalar of mass 270 GeV and its possible connection to the 750 GeV excess

Stefan von Buddenbrock,[a,1] Alan S. Cornell,[b] Deepak Kar,[a] Mukesh Kumar,[b] Bruce Mellado[a] and Robert G. Reed[a]

[a]School of Physics, University of the Witwatersrand, Wits 2050, South Africa
[b]National Institute for Theoretical Physics; School of Physics and Mandelstam Institute for Theoretical Physics, University of the Witwatersrand, Johannesburg, Wits 2050, South Africa

E-mail: [1]stef.von.b@cern.ch

Abstract. Run 1 of the LHC saw many interesting and surprising results, hinting at physics beyond the Standard Model. One of these is the search for lepton pairs in association with high missing energy, which produced a sizeable excess in its rate from both the ATLAS and CMS collaborations. While ATLAS and CMS interpret this using a supersymmetric model, we have shown that the excess can partly be explained by the existence of a new heavy scalar boson, H. This heavy scalar can decay to weak vector bosons and also Standard Model Higgs bosons and missing energy, making it a prime candidate in the search for high p_{T} di-lepton pairs. An analysis is performed to determine the rate of di-lepton production with large missing energy for opposite sign pairs. We show that the heavy scalar produces a non-negligible rate, but cannot fully explain the excesses. For this reason, we look also to incorporating a full two Higgs doublet model to enhance this rate, where the pseudo-scalar A in the spectrum of new particles is considered to be a candidate for the Run 2 di-photon excess at 750 GeV.

1. Key Higgs results from Run 1

By now it is clear that Run 1 of the Large Hadron Collider (LHC) saw the discovery of the Higgs boson, which was measured to have a mass of 125.09 ± 0.24 GeV according to a combination of both ATLAS and CMS results [1]. This particle's couplings have also been shown to behave like a Higgs boson predicted by the Standard Model (SM) [2].

However, studies on the kinematics of this particle have revealed some deviations from SM predictions. In particular, here we can consider the Higgs differential fiducial p_{T} spectra published by both ATLAS [3, 4] and CMS [5, 6]. Looking at the ATLAS results, it is clear that the observed Higgs boson events have significantly more transverse momentum than what is predicted by the SM. Typically, this is indicative of the Higgs boson decaying from some heavy object. The CMS results are not as compelling as the ATLAS results, but the trend is still observable in the $h \to ZZ^* \to 4\ell$ spectrum albeit with a small significance.

Searches for heavy Higgs boson-like particles from Run 1 have also produced some exciting results. For instance, several searches for resonant production of di-Higgs pairs (that is, the decay of some heavy resonance to two Higgs bosons $H \to hh$) from ATLAS [7] and CMS [8, 9, 10] have shown excesses around the ~ 300 GeV region of the resonance mass. A similar effect can be seen in searches for resonant VV production (where V is a vector boson, either Z or $W^{\pm}$). In this channel, CMS [11] sees an excess of events where the resonance mass is about ~ 275 GeV.

This result is compatible with the limits set by ATLAS in the same search channels [12, 13], where a much smaller excess can be seen.

Finally, it is apparent from top associated Higgs production Run 1 results that the SM underestimates the rate of Higgs bosons being produced in association with top quarks. There are many analyses which study this effect, but it can be seen most convincingly when we consider the ATLAS and CMS combination of the μ_{tth} measurement [2]. The parameter μ_{tth} is a signal strength and gives a ratio of the cross section measured for $pp \to t\bar{t}h$ production against what the SM predicts. The measurement lies at

$$\mu_{tth} = \frac{\sigma_{tth}}{\sigma_{tth}^{\rm SM}} = 2.3^{+0.7}_{-0.6},$$

meaning that ATLAS and CMS measure more than two times as many top associated Higgs events than what the SM predicts. It should be noted that all of the other Higgs production mechanisms all have measured signal strengths which are within 1σ of the SM prediction ($\mu = 1$), yet the tth measurement is more than 2σ away. This excess could also be explained by the existence of a heavy Higgs-like particle of which the main decay product is a SM Higgs boson. It has been shown that top associated Higgs production is enhanced if the Higgs boson in question has a small VV branching ratio [14]. The excess, therefore, could be accounted for if we postulate the existence of a heavy scalar H, with a small $H \to VV$ branching ratio and the ability for it to decay to a SM Higgs boson.

2. A 2HDM interpretation

In the section above, it was stated that the excesses and results mentioned can all be explained by proposing the existence of a new heavy scalar H. This approach was taken in reference [15], advocating the existence of a heavy scalar with a mass around ~ 270 GeV using data only from Run 1 of the LHC. The best fit point for this hypothesis was determined to be

$$m_H = 272^{+12}_{-9}\ \text{GeV},$$

and at this mass point the significance of the measurement was of the order of 3σ.

Extending the SM by introducing a single heavy scalar is possible, but a far more natural and theoretically motivated extension is by introducing a second Higgs doublet – often called a two-Higgs doublet model (2HDM) [16]. In this model, the Higgs sector is extended by the addition of all Lorentz invariant terms involving two Higgs doublets, instead of just one as in the SM. The observable consequence of this is the hypothetical existence of four new mass eigenstates in the theory. In addition to the SM Higgs boson h, a 2HDM predicts another CP-even scalar H, a CP-odd scalar A, and two charged scalars $H^{\pm}$.

While the details of 2HDMs are not discussed in this short paper in any detail (a contemporary discussion can be found in reference [17]), it should be noted that the heavy scalar which is the subject of reference [15] could be the extra CP-even scalar found in a 2HDM. If this is the case, then the A and $H^{\pm}$ bosons should also exist in nature. These particles have not been discovered yet, but in this short paper we consider the possibility that the 750 GeV excess in both the ATLAS [18] and CMS [19] di-photon mass spectra from Run 2.

It is important, however, to take note that the heavy scalar scenario presented in reference [15] can not be mapped to a 2HDM alone. In this case, the heavy scalar's production cross section is modified by a factor β_g. Since H is produced dominantly through gluon fusion (ggF), a modification of its production cross section could mean that extra coloured particles run in the ggF loop (in addition to the heavy quarks). If this is the case, then a 2HDM would not completely describe the excesses seen in Run 1. The model also includes a hypothetical dark matter particle, χ, of mass $\sim m_h/2$ which can explain events with large missing transverse energy ($E_{\rm T}^{\rm miss}$).

IOP Conf. Series: Journal of Physics: Conf. Series **802** (2017) 012001 doi:10.1088/1742-6596/802/1/012001

Signal region	Expected	Observed	Significance
Run 1 ATLAS SR-Z [20]	10.6 ± 3.2	29	3.0σ
Run 2 ATLAS SR-Z [21]	10.3 ± 2.3	21	2.2σ
Run 1 CMS low-mass edge-search [22]	730 ± 40	860	2.6σ

Table 1. The event yields for current significant excesses in di-lepton + $E_{\mathrm{T}}^{\mathrm{miss}}$ searches in the literature to date. The Run 2 result from CMS [23] did see a small excess, but this was not classified as significant.

3. Excesses for di-lepton + large $E_{\mathrm{T}}^{\mathrm{miss}}$ events

Several other excesses were observed in Run 1, and some of the most notable of these are in the di-lepton + large $E_{\mathrm{T}}^{\mathrm{miss}}$ channels. These are typically supersymmetry (SUSY) searches, since it is common in SUSY extensions of the SM to expect leptons in the decays of neutralinos – either via a direct decay or from a $Z \to \ell\ell$ decay, since neutralinos also have decay modes involving Z bosons. These searches most commonly look for same-flavour opposite-sign (SFOS) di-lepton pairs, since this is typically what a Z boson can decay to.

The analyses performed for a selection of di-lepton + $E_{\mathrm{T}}^{\mathrm{miss}}$ events are, therefore, largely based on the features predicted by SUSY di-lepton production. Some common selection criteria include:

- Two high p_{T} SFOS leptons: typically seen in the leptonic decay of a Z boson.
- Large missing energy: SUSY events often contain invisible final state particles not detected in the calorimeters.
- Many jets: SUSY models characteristically predict large jet multiplicities in events with di-lepton final states.
- Large H_{T} – that is, the scalar sum of jet transverse momenta: this variable is highly correlated to both jet multiplicity and $E_{\mathrm{T}}^{\mathrm{miss}}$, and has therefore been left out of some signal regions in the literature because of this redundancy.

The different signal regions (SRs) used by ATLAS and CMS make use of different combinations of cuts on the variables mentioned above. They can further be classified as either peak (on-Z) or edge (off-Z) searches, where the former restricts the invariant mass of the di-lepton pair to be close to the Z boson mass and the latter excludes this region. The significant excesses as reported by ATLAS and CMS in their respective SRs are shown in Table 1, which compares the expected and observed rates.

The SUSY models considered in these searches can of course be tuned to match the observed rates, but when one considers the differential distributions of a few key variables in the study it becomes clear that the SUSY models do not explain the excesses well. In this short paper, we consider only the peak search SR from ATLAS (SR-Z) since this is where the excess is largest and the differential distributions from this analysis are clear enough to make a comparison between the SUSY models and the data. One can see from the ATLAS SR-Z jet multiplicity distributions that events passing the selection criteria have far fewer jets than what the SUSY models predict. This is confirmed by the H_{T} distributions. The data seem to show that H_{T} decreases like the tail of a distribution, whereas the SUSY models predict a peak at large H_{T}. These two observations are a good argument against a SUSY explanation of the excess, primarily since the models considered in the publications predict far too many jets.

4. Results in ATLAS SR-Z

In light of the discrepancy mentioned above, it must be mentioned that proposing the existence of the heavy scalar mentioned in section 2 can partly explain the excesses measured while

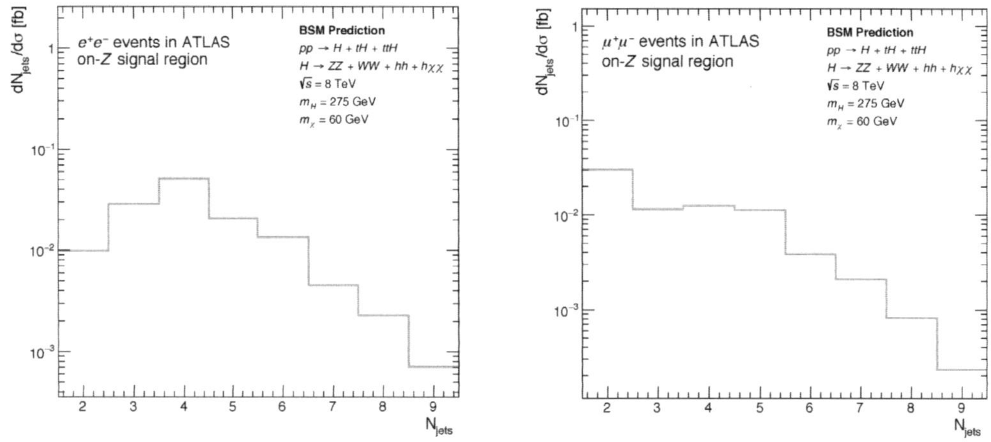

Figure 1. The jet multiplicity differential distributions for the di-lepton signal (e^+e^- on the left and $\mu^+\mu^-$ on the right) generated from heavy scalar production and decays, normalised to a cross section calculable from the parameters in reference [15]. These are intended for comparison against the distributions in reference [20].

predicting differential distributions more closely related to what is seen in the data. For this reason, a simulation was performed to determine whether or not a heavy scalar extension of the SM could match the results given in the ATLAS SR-Z signal region. In detail, the cuts applied for the ATLAS SR-Z signal region are as follows:

- Two SFOS leptons (either e^+e^- or $\mu^+\mu^-$) are selected with a minimum p_{T} of 25(10) GeV for the (sub-)leading lepton.
- The pseudorapidity is bound such that $|\eta| < 2.47(2.4)$ for electrons (muons). For electrons, the crack region is also excluded: $1.37 < |\eta| < 1.52$.
- At least two jets with $p_{\mathrm{T}} > 35$ GeV are required.
- This is a peak-search, so the invariant mass of the di-lepton pair is limited to $81 < m_{\ell\ell} < 101$ GeV.
- The $E_{\mathrm{T}}^{\mathrm{miss}}$ is required to be larger than 225 GeV.
- $H_{\mathrm{T}} = p_{\mathrm{T}}^{\ell 1} + p_{\mathrm{T}}^{\ell 2} + \sum_i p_{\mathrm{T}}^{\mathrm{jet}-i}$ is required to be larger than 600 GeV.
- The azimuthal between the two leading jets four vector and the $E_{\mathrm{T}}^{\mathrm{miss}}$ four vector is required to be larger than 0.4.

Events were generated using `MadGraph` [24] in order to simulate the production of the heavy scalar in its three dominant production modes: $pp \to H$ (ggF), $pp \to tH + \bar{t}H$ (single top associated production) and $pp \to t\bar{t}H$ (double top associated production). These events were then decayed, showered and hadronised in `Pythia 8.2` [25]. The decay modes of H which could generate a signal for the di-lepton + $E_{\mathrm{T}}^{\mathrm{miss}}$ SRs are: $H \to ZZ$ (where $Z \to \ell\ell$ or $\nu\nu$), $H \to WW$ (where $W \to \ell\nu$) and $H \to h\chi\chi$. The last decay mode produces a di-lepton signal in the way the SM Higgs boson decays, and generates large $E_{\mathrm{T}}^{\mathrm{miss}}$ through the existence of the dark matter candidate χ. The mass points considered for this analysis were $m_H = 275$ GeV and $m_\chi = 60$ GeV, which arise from the best fit points in reference [15].

The differential distributions for jet multiplicity from this analysis are shown in Figure 1. It can be seen that the heavy scalar production predicts few jets, especially for the case of a

IOP Conf. Series: Journal of Physics: Conf. Series **802** (2017) 012001 doi:10.1088/1742-6596/802/1/012001

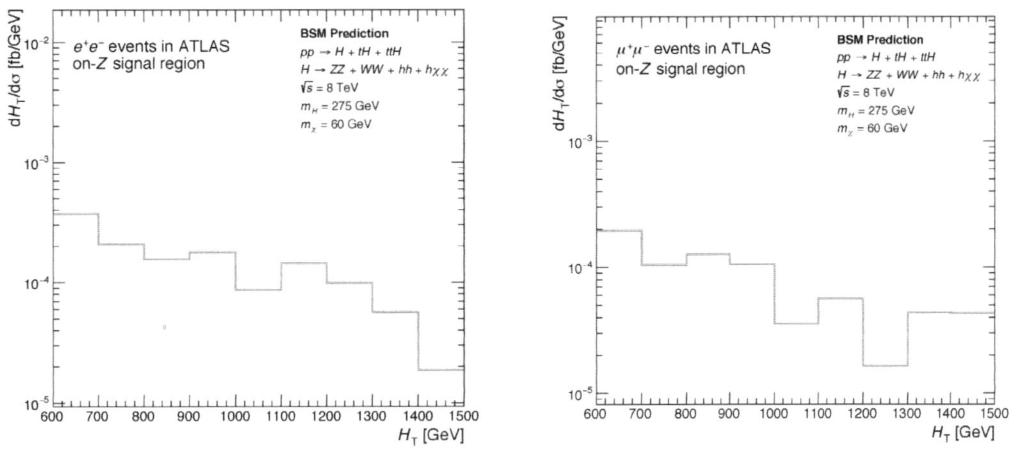

Figure 2. The jet multiplicity differential distributions for the di-lepton signal (e^+e^- on the left and $\mu^+\mu^-$ on the right) generated from heavy scalar production and decays, normalised to a cross section calculable from the parameters in reference [15]. These are intended for comparison against the distributions in reference [20].

$\mu^+\mu^-$ pair. This is confirmed by the H_T differential distributions in Figure 2, which show what appears to be the tail of the distributions. When one compares these two distributions to the ATLAS data, it is apparent that they explain the shapes of what is observed fairly well, and far more convincingly than the SUSY models considered in the published analyses.

The rate, however, does not fully explain the large excesses observed. Using the constrained parameters in reference [15], a total rate was able to be predicted for the production mechanisms and decays mentioned above. The prediction is a total of 4.2±1.8 events. This is a non-negligible rate, but is clearly not enough to saturate the $\sim$ 18 event excess seen in the ATLAS data. This is where the need for the CP-odd Higgs boson A comes about.

Hypothetically, the rate of di-lepton events with large E_T^{miss} will be greatly enhanced by the production of the A pseudo-scalar. While there is not enough information about the model as yet to predict a rate, a study was done to calculate the fiducial efficiency of events which could pass the selection mentioned above, through the production $pp \to A \to ZH$ (where $H \to ZZ$, WW, $h\chi\chi$). When the mass points $m_A = 750$ GeV, $m_H = 275$ GeV and $m_\chi = 60$ GeV are chosen, the selection efficiency was of the order of 20%, which is large and shows promise for a sizeable rate in the di-lepton + E_T^{miss} search channel as well as an explanation for the 750 GeV di-photon excesses.

5. Conclusions

A number of unexplained excesses were seen in Run 1 of the LHC, both from ATLAS and CMS. The proposition has been made that many of these excesses can be explained by postulating the existence of a heavy scalar with mass around 270 GeV (as shown in reference [15]). When applying this hypothesis to excesses seen in di-lepton + E_T^{miss} search channels, it is observed that the heavy scalar can explain part of the excess as well as the shapes of the differential distributions seen in data. The rate predicted, however, is not comparable with the large excess observed by ATLAS in Run 1. For this reason, a small study has been done to determine how effective a pseudo-scalar at 750 GeV could be at enhancing this rate. This pseudo-scalar was postulated as an explanation for the 750 GeV excess seen in the ATLAS and CMS Run 2 di-photon mass spectra. The efficiency for di-lepton + $E_{\mathrm{T}}^{\mathrm{miss}}$ events coming from the process $pp \to A \to ZH$ is a promising large number, and requires further study to determine whether its existence could help saturate the large excess seen in Run 1.

References

[1] Aad G *et al.* (ATLAS, CMS) 2015 *Phys. Rev. Lett.* **114** 191803 (arXiv: 1503.07589)
[2] Aad G *et al.* (ATLAS, CMS) 2015 ATLAS-CONF-2015-044
[3] Aad G *et al.* (ATLAS) 2014 *JHEP* **09** 112 (arXiv: 1407.4222)
[4] Aad G *et al.* (ATLAS) 2014 *Phys. Lett.* **B738** 234–253 (arXiv: 1408.3226)
[5] CMS Collaboration 2015 CMS-PAS-HIG-14-028
[6] Khachatryan V *et al.* (CMS) 2016 *Eur. Phys. J.* **C76** 13 (arXiv: 1508.07819)
[7] Aad G *et al.* (ATLAS) 2015 *Phys. Rev.* **D92** 092004 (arXiv: 1509.04670)
[8] CMS Collaboration 2014 CMS-PAS-HIG-13-032
[9] Khachatryan V *et al.* (CMS) 2015 (arXiv: 1510.01181)
[10] Khachatryan V *et al.* (CMS) 2014 *Phys. Rev.* **D90** 112013 (arXiv: 1410.2751)
[11] Khachatryan V *et al.* (CMS) 2015 *JHEP* **10** 144 (arXiv: 1504.00936)
[12] Aad G *et al.* (ATLAS) 2016 *Eur. Phys. J.* **C76** 45 (arXiv: 1507.05930)
[13] Aad G *et al.* (ATLAS) 2016 *JHEP* **01** 032 (arXiv: 1509.00389)
[14] Farina M, Grojean C, Maltoni F, Salvioni E and Thamm A 2013 *JHEP* **05** 022 (arXiv: 1211.3736)
[15] von Buddenbrock S, Chakrabarty N, Cornell A S, Kar D, Kumar M, Mandal T, Mellado B, Mukhopadhyaya B and Reed R G 2015 (arXiv: 1506.00612)
[16] Lee T D 1973 *Phys. Rev.* **D8** 1226–1239 [,516(1973)]
[17] Branco G C, Ferreira P M, Lavoura L, Rebelo M N, Sher M and Silva J P 2012 *Phys. Rept.* **516** 1–102 (arXiv: 1106.0034)
[18] ATLAS collaboration 2015 ATLAS-CONF-2015-081
[19] CMS Collaboration 2015 CMS-PAS-EXO-15-004
[20] Aad G *et al.* (ATLAS) 2015 *Eur. Phys. J.* **C75** 318 [Erratum: Eur. Phys. J.C75,no.10,463(2015)] (arXiv: 1503.03290)
[21] ATLAS collaboration 2015 ATLAS-CONF-2015-082
[22] Khachatryan V *et al.* (CMS) 2015 *JHEP* **04** 124 (arXiv: 1502.06031)
[23] CMS Collaboration (CMS) 2015 CMS-PAS-SUS-15-011
[24] Alwall J, Herquet M, Maltoni F, Mattelaer O and Stelzer T 2011 *JHEP* **06** 128 (arXiv: 1106.0522)
[25] Sjostrand T, Mrenna S and Skands P Z 2008 *Comput. Phys. Commun.* **178** 852–867 (arXiv: 0710.3820)

Absorption probabilities associated with spin-3/2 particles near N-dimensional Schwarzschild black holes

G. E. Harmsen[a,1] **C. H. Chen**[b,2], **H. T. Cho**[b,3], **and A. S. Cornell**[a,4]

[a] National Institute for Theoretical Physics; School of Physics and Mandelstam Institute for Theoretical Physics, University of the Witwatersrand, Johannesburg, Wits 2050, South Africa
[b] Department of Physics, Tamkang University, Tamsui, Taipei, Taiwan

E-mail: [1]gerhard.harmnen5@gmail.com, [2]899210057@s99.tku.edu.tw, [3]htcho@mail.tku.edu.tw, [4]alan.cornell@wits.ac.za

Abstract. In June 2015 the Large Hadron Collider was able to produce collisions with an energy of 13TeV, where collisions at these energy levels may allow for the formation of higher dimensional black holes. In order to detect these higher dimensional black holes we require an understanding of their emission spectra. One way of determining this is by looking at the absorption probabilities associated with the black hole. In this proceedings we will look at the absorption probability for spin-3/2 particles near N-dimensional Schwarzschild black holes. We will show how the Unruh method is used to determine these probabilities for low energy particles. We then use the Wentzel-Kramers-Brillouin approximation in order to determine these absorption probabilities for the entire possible energy range.

1. Introduction

If we consider a system containing only a single Schwarzschild black hole and a particle of energy ω, such that ω is very small compared to the total energy of the black hole, then classically the Schwarzschild metric would sufficiently describe the system (as the particle can be considered to be point like). Using only the Schwarzschild metric we could then determine the probability that the particle is absorbed by the black hole. This probability, however, is vastly different to the one that is obtained using a field theoretic approach [1]. In 1976 Unruh showed that even for low energy particles there is significant deviation in absorption probabilities for the classical case and the field theory case [2]. In field theoretic models it is possible to have the particle reflected off the horizon of the black hole. This reflection is interesting as it is equivalent to having a particle escape from the black hole [3], as would be seen with Hawking radiation. In most field theories these reflected, or escaping, particles have associated with them a spectrum of Quasi Normal Modes (QNMs) which are unique to the particle and the black hole from which it was emitted. This QNM spectra can be used to uniquely describe the parameters of the black hole from which the particle was emitted. A detailed description of QNMs and how they can be calculated is given in Refs. [4, 5].

This proceedings will focus on the absorption probability of spin-3/2 particles, where our motivation for studying this particle is two fold: Firstly from a theoretical point, due to the scarcity of work focusing on these particles. Secondly, in many super gravity theories gravity is strongly coupled to the massless Rarita-Schwinger fields, spin-3/2 fields, which act as a source of torsion and curvature on the space time [6, 7].

Using the method developed by Unruh we will present the equations required to determine the absorption probabilities for the low energy cases [2]. We will then use the WKB method developed by Iyer and Will, to determine a solution for the more general case [8].

2. N-dimensional Schwarzschild

2.1. *Potential function*

In order to implement the Unruh method we must use a Klein-Gordon equation to describe our particle. We have chosen to use the massless Rarita-Schwinger equation,

$$\gamma^{\mu\nu\alpha}\nabla_\nu \Psi_\alpha = 0 \,, \tag{1}$$

where $\gamma^{\mu\nu\alpha} = \gamma^\mu\gamma^\nu\gamma^\alpha + \gamma^\mu g^{\nu\alpha} - \gamma^\nu g^{\mu\alpha} + \gamma^\alpha g^{\mu\nu}$. We also need to describe the space time this is being calculated in using the N-dimensional metric,

$$ds^2 = -f(r)dt^2 + f(r)^{-1}dr^2 + r^2 d\Omega^2_{N-2} \,, \tag{2}$$

where $d\Omega^2_{N-2} = d\theta^2 + \sin\theta d\theta^2_{N-3}$ is the metric describing the $N-2$ sphere. Using Eq.(1) and Eq.(2) we can derive our equations of motion.

In the case of N-dimensional black holes we have two sets of equations of motion, and therefore two potentials. This is because we have both spinors and spinor-vectors in the N-dimensional space time, spinor-vectors have both TT-eigenmodes and non-TT eigenmodes. Please refer to Ref. [9] for a complete description. In the interests of brevity we will focus only on the spinor eigenmodes as the method and results are similar for the spinor-vector eigenmodes.

In solving for our potential we begin by determining the equations of motion of our particle. We then perform a conformal transformation on these equations of motion to obtain a Schrödinger like equation, which we have called a radial equation. From this radial equation we can determine our potential function. The radial equations are given as,

$$-\frac{d^2}{dr_*^2}\tilde{\Phi}_1 + V_1\tilde{\Phi}_1 = \omega^2\tilde{\Phi}_1 \quad , \quad -\frac{d^2}{dr_*^2}\tilde{\Phi}_2 + V_2\tilde{\Phi}_2 = \omega^2\tilde{\Phi}_2 \,. \tag{3}$$

where r_* is the tortoise coordinate, given as $d/dr_* = f(r)d/dr$ and $f(r) = 1 - (2M/r)^{N-3}$, and our potential is [9]

$$V_{1,2} = \pm f\frac{dW}{dr} + W^2, \tag{4}$$

with

$$W = \frac{|\bar{\lambda}|\sqrt{f}}{r}\left(\frac{\left(\frac{2}{N-2}\right)^2|\bar{\lambda}|^2 - 1 - \frac{N-4}{N-2}\left(\frac{2M}{r}\right)^{N-3}}{\left(\frac{2}{N-2}\right)^2|\bar{\lambda}|^2 - f}\right). \tag{5}$$

In solving for the absorption probabilities we will work with V_1 and set $M = 1$.

2.2. *The Unruh method*

There are three regions around the black hole that we must consider. Namely the near region, where $f(r) \to 0$, the central region, where $V(r) \gg \omega$, and the far region, where $f(r) \to 1$. Approximations are obtained for each of the regions and then coefficients are determined by comparing and evaluating the solutions at the boundaries.

IOP Conf. Series: Journal of Physics: Conf. Series **802** (2017) 012002 doi:10.1088/1742-6596/802/1/012002

2.2.1. Near region In the near region $f(r) \to 0$, Eq.(3) becomes

$$\left(\frac{d^2}{dr_*^2} + \omega^2\right) \widetilde{\Phi}_I = 0\,, \tag{6}$$

which has a solution

$$\widetilde{\Phi}_I = A_I e^{i\omega r_*}. \tag{7}$$

2.2.2. Central region In this region we have that $V(r) \gg \omega$ and hence Eq.(3) becomes

$$\left(\frac{d}{dr_*} + W\right)\left(\frac{d}{dr_*} - W\right) \widetilde{\Phi}_{II} = 0\,. \tag{8}$$

Defining H as

$$H = \left(\frac{d}{dr_*} - W\right) \widetilde{\Phi}_{II}\,, \tag{9}$$

the solution of Eq.(9) is

$$H = B_{II}\left(\frac{1+\sqrt{f}}{1-\sqrt{f}}\right)^{\frac{j}{N-3}+\frac{1}{2}} \left(\frac{\left(\frac{2}{N-2}\right)\left(j+\frac{N-3}{2}\right) - \sqrt{f}}{\left(\frac{2}{N-2}\right)\left(j+\frac{N-3}{2}\right) + \sqrt{f}}\right). \tag{10}$$

Substituting Eq.(10) into Eq.(9), we have a first order differential equation with solution,

$$\widetilde{\Phi}_{II} = A_{II}\left(\frac{1+\sqrt{f}}{1-\sqrt{f}}\right)^{\frac{j}{N-3}+\frac{1}{2}} \left(\frac{\left(\frac{2}{N-2}\right)\left(j+\frac{N-3}{2}\right) - \sqrt{f}}{\left(\frac{2}{N-2}\right)\left(j+\frac{N-3}{2}\right)} + \sqrt{f}\right) + B_{II}\Psi\,, \tag{11}$$

where Ψ is

$$\begin{aligned}\Psi = &\left(\frac{1+\sqrt{f}}{1-\sqrt{f}}\right)^{\frac{j}{N-3}+\frac{1}{2}} \left(\frac{\left(\frac{2}{N-2}\right)\left(j+\frac{N-3}{2}\right) - \sqrt{f}}{\left(\frac{2}{N-2}\right)\left(j+\frac{N-3}{2}\right)} + \sqrt{f}\right) \\ &\times \left[\int^{r} \frac{1}{f}\left(\frac{1-\sqrt{f}}{1+\sqrt{f}}\right)^{\frac{2j}{N-3}+1} \left(\frac{\left(\frac{2}{N-2}\right)\left(j+\frac{N-3}{2}\right) + \sqrt{f}}{\left(\frac{2}{N-2}\right)\left(j+\frac{N-3}{2}\right)} - \sqrt{f}\right)^{2} dr'\right].\end{aligned} \tag{12}$$

2.2.3. Far region In the far region $f(r) \to 1$, Eq.(3) becomes

$$\frac{d^2}{dr_*^2}\widetilde{\Phi}_{III} - \left[\frac{\left(\left(j+\frac{D-4}{2}\right)^2 - \frac{1}{4}\right)}{r^2} - \omega^2\right] \widetilde{\Phi}_{III} = 0\,. \tag{13}$$

In this region $r_* \sim r$, the solution can be expressed as a Bessel function

$$\widetilde{\Phi}_{III} = A_{III}\sqrt{r}J_{j+\frac{N-4}{2}}(\omega r) + B_{III}\sqrt{r}N_{j+\frac{N-4}{2}}(\omega r)\,. \tag{14}$$

Setting the amplitude of $\widetilde{\Phi}_{III}$ to one at spatial infinity, comparing the solutions of region I and II, and regions II and III, we find that the probability amplitude is

$$|A_j(\omega)|^2 = 4\pi C^2 \omega^{2j+N-3}\left(1 + \pi C^2\omega^{2j+N-3}\right)^{-2} \approx 4\pi C^2\omega^{2j+N-3}\,. \tag{15}$$

The constant

$$C = \frac{1}{2^{\frac{N-1}{N-3}j+\frac{N-1}{2}}\Gamma(j+\frac{N-2}{2})}\begin{pmatrix} j+\frac{2N-5}{2} \\ j-\frac{1}{2}\end{pmatrix}, \tag{16}$$

with Γ denoting the gamma function and ω being less than 1. This restriction on ω means that we require another method in order to determine the absorption probability of our entire energy spectrum of particles. Schutz and Clifford developed a semi-analytical technique, based on the WKB method, that can accurately reproduce the numerical values for QNMs obtained using other methods [10]. We use this variant of the WKB method to solve for our more energetic particles.

2.3. WKB method

It is more convenient in this case to write $x = \omega r$ and take $Q(x) = \omega^2 - V$. Eq.(3) then becomes

$$\left(\frac{d^2}{dr_*^2} + Q\right)\widetilde{\Phi}_1 = 0 . \tag{17}$$

For energies $\omega \ll V$ it is sufficient to use the first order WKB approximation [11],

$$|A_j| = exp\left[-2\int_{x_1}^{x_2}\frac{dx'}{f(x')}\sqrt{-Q(x')}\right], \tag{18}$$

where x_1 and x_2 are turning points, with $Q(x_1, x_2) = 0$ or $V_{x_1,x_2} = \omega^2$, for a given energy ω and potential V. When $\omega^2 \sim V$ the formula in Eq.(18) is no longer convergent. For this energy we will need to use a higher order WKB approximation. We use the third order method developed by Iyer and Will where they have shown that the absorption probability can be written as

$$|A_j(\omega)|^2 = \frac{1}{1+e^{2S(\omega)}}, \tag{19}$$

where $S(\omega)$ is given as

$$\begin{aligned} S(\omega) = &-\pi k^{-1/2}\left[\frac{1365}{2048}b_3^4 - \frac{525}{256}b_3^2 b_4 + \frac{85}{128}b_4^2 + \frac{85}{128}b_4^2 + \frac{95}{64}b_3 b_5 - \frac{25}{32}b_6\right] z_0^2 \\ &+ \pi k^{-1/2}\left[\frac{3}{16}b_4 - \frac{7}{64}b_3^2\right] + \pi k^{1/2}\left[\frac{1}{2}z_0^2 + \left(\frac{15}{64}b_3^2 - \frac{3}{16}b_4\right)z_0^4\right] \\ &+ \pi k^{1/2}\left[\frac{1155}{2048}b_3^4 - \frac{315}{256}b_3^2 b_4 + \frac{35}{128}b_4^2 + \frac{35}{64}b_3 b_5 - \frac{5}{32}b_6\right] z_0^6 , \end{aligned} \tag{20}$$

for z_0^2, b_n and k defined by the components of the Taylor series expansion of $Q(x)$ near x_0. That is,

$$Q = Q_0 + \frac{1}{2}Q_0''z^2 + \sum_{n=3}\frac{1}{n!}\left(\frac{d^nQ}{dx^n}\right)_0 z^n = k\left[z^2 - z_0^2 + \sum_{n=3} b_n z^n\right], \tag{21}$$

and

$$z = x - x_0;\ z_0^2 \equiv -2\frac{Q_0}{Q_0''};\ k \equiv \frac{1}{2}Q_0'';\ b_n \equiv \left(\frac{2}{n!Q_0''}\right)\left(\frac{d^nQ}{dx^n}\right)_0, \tag{22}$$

where subscript 0 denotes the maximum and the primes denote derivatives.

IOP Conf. Series: Journal of Physics: Conf. Series **802** (2017) 012002 doi:10.1088/1742-6596/802/1/012002

3. Results and conclusions

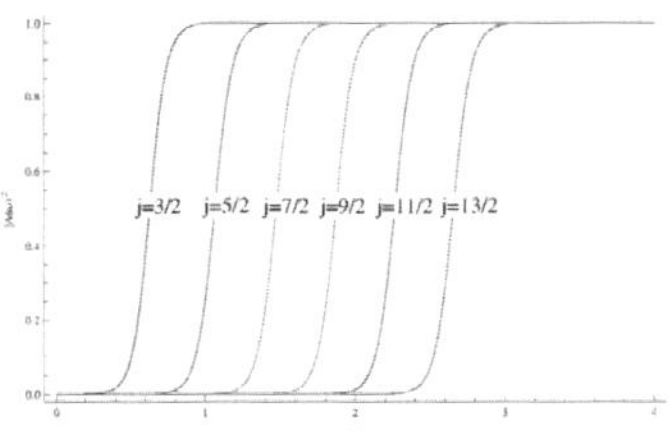

(a) Absorption probability and particle energy for 4-dimensional Schwarzschild black hole.

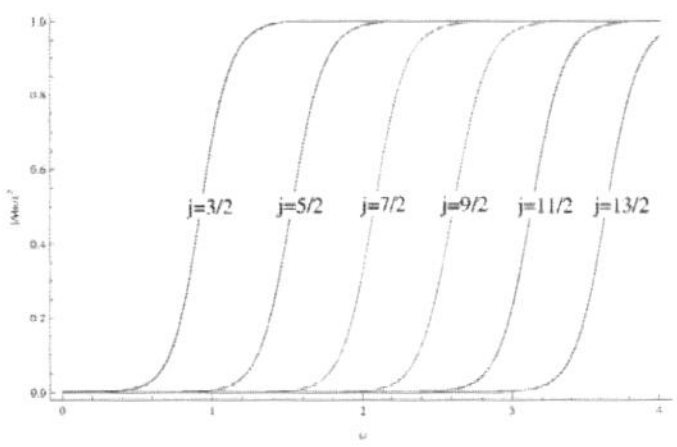

(b) Absorption probability and particle energy for 5-dimensional Schwarzschild black hole.

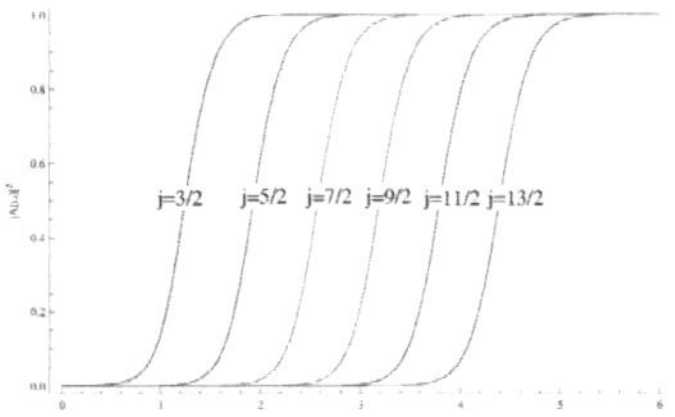

(c) Absorption probability and particle energy for 6-dimensional Schwarzschild black hole.

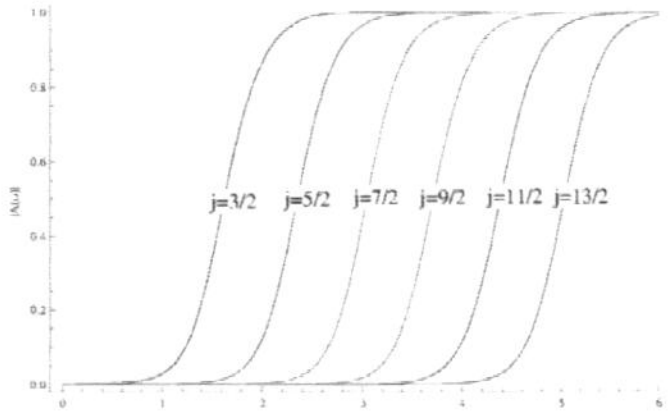

(d) Absorption probability and particle energy for 7-dimensional Schwarzschild black hole.

Figure 1: Absorption probabilities associated to particles of spin-3/2 with j values ranging from 3/2 to 13/2 near Schwarzschild black holes of dimension 4 to 7.

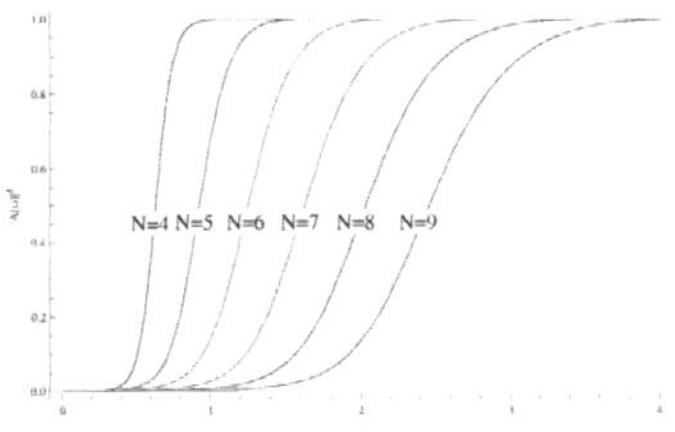

(a) Absorption probability and particle energy for j=3/2 on various dimensional Schwarzschild black holes.

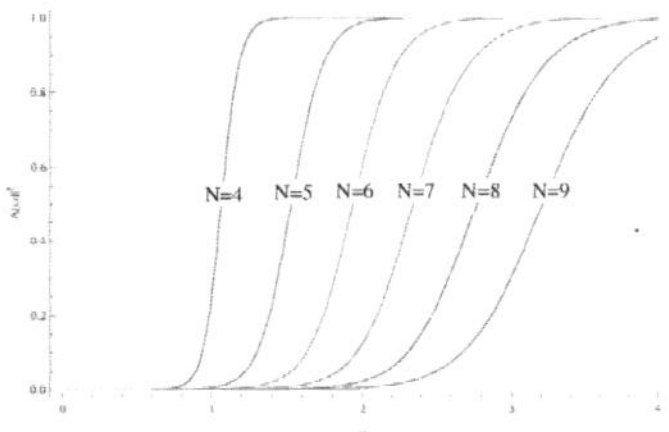

(b) Absorption probability and particle energy for j=5/2 on various dimensional Schwarzschild black holes.

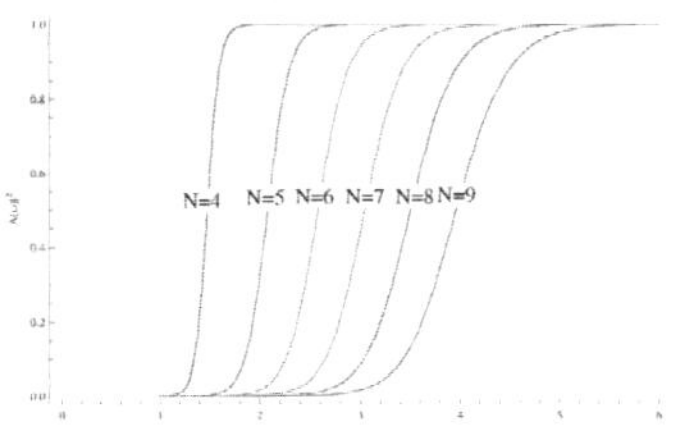

(c) Absorption probability and particle energy for j=7/2 on various dimensional Schwarzschild black holes.

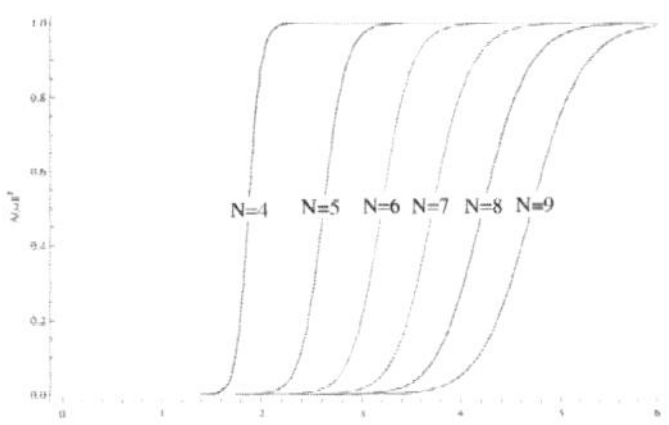

(d) Absorption probability and particle energy for j=9/2 on various dimensional Schwarzschild black holes.

Figure 2: Absorption probability for particles of $j = 3/2$ to $j = 9/2$ for various dimensional Schwarzschild black holes.

The results for our absorption probabilities, for spinors, are given in Figs. (1) and (2). Although we have not explicitly given the effective potential for the spinor-vectors, they are derived in the same way as the potential for our spinors. It is interesting to note that the potential for our spinor-vectors can be reduced to the potential for spin-1/2 particles [9, 12]. Therefore if we look at the absorption probability for spin-1/2 particles and our spin-3/2 spinor vector particles we find that their absorption probabilities are consistent with each other.

Setting $N = 4$ in our potential yields the same potential as we have obtained in Ref. [13]. Looking at Figs. (1a) to (1d) we see that increasing the excitation level of our particle increases the minimum energy required for absorption to take place. This means that more excited particles will be emitted from the black hole with a higher energy. Next, looking at Figs. (2a) and (2b) we see that higher dimensional black holes will emit more energetic particles, since the minimum energy required for absorption increases with an increasing number of dimensions.

Acknowledgements

We would like thank Wade Naylor for his useful discussions during the production of this work. ASC and GEH are supported in part by the National Research Foundation of South Africa (Grant No: 91549). CHC and HTC are supported in part by the Ministry of Science and Technology, Taiwan, ROC under the Grant No. NSC102-2112-M-032-002-MY3, and by the National Center for Theoretical Sciences (NCTS). GEH would like to thank CHC and HTC for their hospitality during his visit to Tamkang university, where part of this work was completed.

References

[1] Funkhouser S 2010 Particle absorption by black holes and the generalized second law of thermodynamics *Proceedings of the Royal Society of London A: Mathematical, Physical and Engineering Sciences* vol 466 (The Royal Society) pp 1155–1166
[2] Unruh W 1976 *Physical Review D* **14** 3251
[3] Kuchiev M Y 2004 *EPL (Europhysics Letters)* **65** 445
[4] Konoplya R and Zhidenko A 2011 *Reviews of Modern Physics* **83** 793
[5] Kokkotas K D and Schmidt B G 1999 *Living Rev. Rel* **2** 262
[6] Das A and Freedman D Z 1976 *Nuclear Physics B* **114** 271–296
[7] Grisaru M T, Pendleton H and Van Nieuwenhuizen P 1977 *Physical Review D* **15** 996
[8] Iyer S and Will C M 1987 *Physical Review D* **35** 3621
[9] Chen C H, Cho H T, Cornell A S and Harmsen G E 2016 *WITS-MITP-022*
[10] Schutz B F and Will C M 1985 *The Astrophysical Journal* **291** L33–L36
[11] Cho H T and Lin Y 2005 *Classical and Quantum Gravity* **22** 775
[12] Cho H T, Cornell A S, Doukas J and Naylor W 2007 *Physical Review D* **75** 104005
[13] Chen C H, Cho H T, Cornell A S, Harmsen G and Naylor W 2015 *Chinese Journal Of Physics* **53**

An Extremely Brief Introduction to Quantum Field Theory

W. A. Horowitz

Department of Physics, University of Cape Town, Private Bag X3, Rondebosch 7701, South Africa

E-mail: wa.horowitz@uct.ac.za

Abstract. We provide a very brief introduction to n-dimensional scalar field theory, with an eye to renormalization and expectation values of operators. We assume the audience already has some experience with QFT.

1. Introduction: Making Quantum Mechanics Relativistic

Recall the postulates of non-relativistic quantum mechanics (NRQM):

(i) the state of the system is represented by a vector $|\psi\rangle \in \mathcal{H}$, a Hilbert space

(ii) observables are represented by (essentially) self-adjoint operators that obey the Dirac quantization condition, $\underbrace{\{A, B\}_{classical}}_{Poisson\ bracket} = C \to [\hat{A}, \hat{B}]_{commutator} = i\hat{C}$

(iii) if the system is in a state $|\psi\rangle$, then the measurement of an observable represented by $\hat{\mathcal{O}}$ will be an eigenvalue ω of $\hat{\mathcal{O}}$, $\hat{\mathcal{O}}|\omega\rangle = \omega|\omega\rangle$ with probability $P(\omega) \propto |\langle\omega|\psi\rangle|^2$; as a result of the measurement, the state of the system changes instantaneously from $|\psi\rangle$ to $|\omega\rangle$

(iv) $|\psi(t)\rangle$ evolves according to the Schrödinger equation, $i\partial_t|\psi(t)\rangle = \hat{H}(t)|\psi(t)\rangle$, where $\hat{H}$ is the Hamiltonian operator for the system, which may explicitly (or implicitly) depend on time t; we usually take $\hat{H} = \hat{p}^2/2m + V(\hat{x})$.

Now recall the postulates of special relativity (SR):

(i) the laws of physics are the same in all inertial reference frames (inertial frames move with constant velocity with respect to each other; observers in different frames must agree on the results of physical measurements; and the laws of physics must be the same, which is to say that all inertial observers agree on the initial conditions and equations of motion for the system)

(ii) the speed of light is a constant $c = 1$ in all inertial frames, which is equivalent to requiring that our physical models are on a pseudo-Riemannian manifold with metric $\eta^{\mu\nu}$ with diagonal elements $\{+1, -1, -1, -1, \ldots\}$, which is again equivalent to requiring that space and time are treated on equal footing; physical quantities transform under the Lorentz group, for example $U(\Lambda)\phi(x) = \phi(\Lambda^{-1}x)$ for a scalar field; finally, the speed of light acts as a speed limit—information travels (at most) as fast as the speed of light in vacuum.

Notice that some of the postulates of NRQM appear to conflict with the postulates of SR. In particular, wavefunction collapse in NRQM is immediate and the Schrödinger equation is not relativistically invariant. The issue of instantaneous wavefunction collapse was first raised by Einstein, Podolsky, and Rosen [1]. We will avoid the issue in two ways: first, wavefunction collapse transmits no information (and therefore does not violate the speed limit imposed by the finite speed of light); and second, in quantum field theory (QFT), we only ask questions about asymptotic past or future states of the system.

The Schrödinger equation breaks Lorentz invariance in two ways. First, the usual Hamiltonian $\hat{p}^2/2m + V(\hat{x})$ does not yield the relativistically correct dispersion relation. One may readily modify the Hamiltonain to $\hat{H} = \sqrt{\hat{p}^2 + m^2}$, leading to the correct relation $E = \sqrt{p^2 + m^2}$. However, the Schrödinger equation breaks Lorentz invariance in a second, fundamental way by treating time and space as fundamentally different: the Schrödinger equation is linear in ∂_t but quadratic in ∂_x; similarly, position is elevated to an operator while time is only a parameter.

In order to resolve the conflict between NRQM and SR we have two options: promote t to an operator (the string theory like approach [2, 3]) or demote $\hat{x}$ to a parameter and treat fields as operators (particle theory like approach [4–12]). As this is a particle physics context, we will follow the latter approach.

2. The Relativistic Quantum Scalar Field

We will take symmetries as the fundamental organizing principle for QFT.

In path integral language, we seek a Lagrangian that respects the symmetries we postulate for our physics. (Path integrals perfectly preserve classically the symmetries of the Lagrangian explicitly. One finds that quantum mechanics leads to anomalies: renormalization can break classical symmetries.)

Let's start by postulating that scalar particles are the quanta of a field ϕ whose Lagrangian is

$$\mathcal{L} = \frac{1}{2}(\partial_\mu \phi_0)^2 - \frac{1}{2} m_0^2 \phi_0^2 + \lambda_0 \phi_0^4. \tag{1}$$

Scalars are valuable because they are the simplest mathematical objects to work with (c.f. fermions, vector fields, etc.). Scalars also describe a wealth of physics (e.g. the Higg's boson and scalar and pseudoscalar mesons).

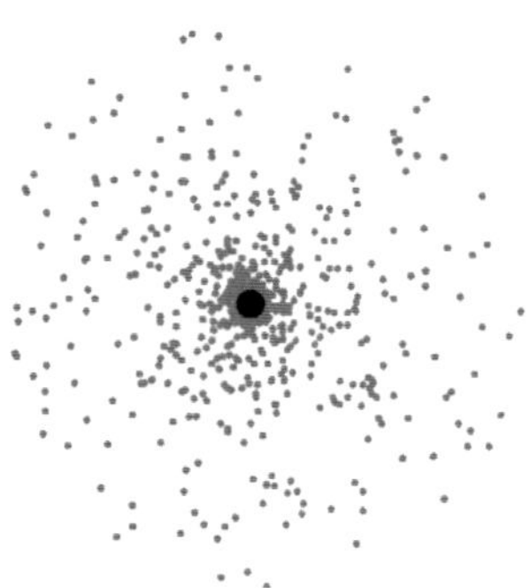

Figure 1. The cloud of interacting particles (blue) surrounding the "original" particle p (black). The particles are actually all identical, so the different colors and sizes for the particles is for visualization purposes only.

Eq. 1 is written in terms of bare quantities ϕ_0, m_0, and λ_0 in anticipation of performing renormalized perturbation theory, in which we will re-write the Lagrangian in terms of renormalized quantities ϕ_r, m_r, and λ_r.

Renormalization is a generic feature of QFT (not a bug!). Mathematically, there are infinities that need to be rigorously dealt with. Physically, we require that the parameters in our theory are connected in some way to a physical measurement. Additionally, our theories will only make sense if the physics at one scale is not affected by the physics at a dramatically different scale. (For example, the physics at infinitesimally small distances shouldn't alter the dynamics at macroscopic lengths.)

Remember that we are dealing with an inherently quantum mechanical system, which implies Heisenberg's Uncertainty Principle, which implies that virtual pair production is possible; i.e. particle–anti-particle pairs pop in and out of existence all the time. Our theory is self-interacting. So for any one physical particle p of our theory, there is a cloud of virtual particles surrounding it, all interacting with each other and with p; see Fig. 1. In fact, physically, we cannot tell the difference between p and its cloud. So what an experimentalist measures for the charge of p (related to λ_0) or the mass of p (related to m_0) or even how many p's there are (related to ϕ_0) may depend on the energy scale (or, equivalently, distance) at which they probe p.

3. Quantizing the Free Relativistic Scalar Field

Our goal is to quantize the free relativistic scalar field in n spatial dimensions. (Note that we will use a more naturally relativistic notation than Peskin and Schroeder [7], although we will keep the particle physicist's mostly minus convention for the metric.)

We will postulate that the relevant relativistically invariant Lagrange (density) is

$$\mathcal{L} = \frac{1}{2}(\partial_\mu \hat{\phi}_0)^2 - \frac{1}{2} m_0^2 \hat{\phi}_0^2. \tag{2}$$

We seek a solution $\hat{\phi}_0$ associated with the above Lagrange (density) and a spectrum of the field operator (eigenvalues and eigenvectors). We will find that $\hat{\phi}_0$ is the bare (unrenormalized) operator associated with the production of a single scalar particle of mass m_0. For notational simplicity, for the rest of this section we will drop the 0 subscript and the operator hat. (Note that since the theory is non-interacting, we can solve the theory exactly and without renormalization.)

Dimensional analysis is an invaluable tool for the physicist, and so we should pause for a moment to analyze the dimensions of the objects in our theory. The path integral goes as $\int \mathcal{D}\phi e^{iS}$, where $S = \int dx^{n+1}\mathcal{L}$. Since one can only take a number to the power of a dimensionless number, the action must be dimensionless, $[S] = 1$. Since $[d^{n+1}x] = L^{n+1} = E^{-(n+1)}$, we have that $[\mathcal{L}] = E^{n+1}$. Since $[\partial_\mu] = [m] = E$, we have that $[\phi] = E^{(n-1)/2}$.

Our solution ϕ must first of all satisfy the classical equations of motion found by extremizing the action. The Euler-Lagrange equations yield

$$(\Box + m^2)\phi = 0. \tag{3}$$

Already things are looking very promising: the equations of motion are Lorentz invariant (as they must be as the Lagrangian is Lorentz invariant).

We may readily solve Eq. 3 by decomposing our solution into Fourier modes:

$$\hat{\phi}(x) = \int \frac{d^n p}{(2\pi)^n 2E_{\vec{p}}} \left(\hat{a}_{\vec{p}} e^{-ip\cdot x} + \hat{a}_{\vec{p}}^\dagger e^{ip\cdot x}\right)_{p^0 = E_{\vec{p}} = \sqrt{\vec{p}^2 + m^2}}. \tag{4}$$

Notice how: 1) we have (explicitly) separated out the classical from the quantum in Eq. 4, where $\hat{\phi}^\dagger = \hat{\phi}$ automatically; 2) the Fourier modes obey the usual relativistic dispersion relation; and 3) dimensional analysis implies that $E^{(n-1)/2} = E^{n-1}[\hat{a}]$, so then $[\hat{a}] = E^{(1-n)/2} = [\hat{a}^\dagger]$.

In order to quantize Eq. 4, we must impose the Dirac quantization condition. We must of course then first decide what the Dirac quantization condition is. Since we are interested in relativistic theories, we will require that fields cannot influence each other outside of the lightcone. Hence a sensible generalization from the 1D NRQM case of $[\hat{x}, \hat{p}] = i$ for the Dirac quantization condition for fields is to require an equal-time contact interaction:

$$[\hat{\phi}(x^0, \vec{x}), \hat{\pi}(x^0, \vec{y})] = i\,\delta^{(n)}(\vec{x} - \vec{y}); \tag{5}$$

i.e. fields at the same time (in one inertial frame) may only affect each other at exactly equal points (otherwise information could propagate faster than the speed of light).

As an aside, it's an interesting exercise, once we've solved for $\phi(x)$, to compute the commutator $[\hat{\phi}(x), \hat{\phi}(y)]$ without the equal time restriction.

In order to impose our proposed commutator relation Eq. 5, we need to compute the canonically conjugate momentum to the field: $\pi(x) \equiv \partial\mathcal{L}/\partial(\partial_0\phi) = \partial^0\phi = \partial_0\phi$. One may easily see that

$$\hat{\pi}(x) = -\frac{i}{2}\int\frac{d^np}{(2\pi)^n}\left(\hat{a}_{\vec{p}}e^{-ip\cdot x} - \hat{a}^\dagger_{\vec{p}}e^{ip\cdot x}\right)_{p^0=E_{\vec{p}}=\sqrt{\vec{p}^2+m^2}}. \tag{6}$$

One may then compute the equal-time commutator

$$[\hat{\phi}(x^0,\vec{x}),\hat{\pi}(x^0,\vec{y})] = \int\frac{d^npd^nq}{(2\pi)^n(2\pi)^n}\frac{-i}{4E_{\vec{p}}}\left\{-[\hat{a}_{\vec{p}},\hat{a}^\dagger_{\vec{q}}]e^{i(q\cdot y-p\cdot x)} + [\hat{a}^\dagger_{\vec{p}},\hat{a}_{\vec{q}}]e^{i(p\cdot x-q\cdot y)}\right\}. \tag{7}$$

We postulated that Eq. 7 should equal $i\delta^{(n)}(\vec{x}-\vec{y})$, Eq. 5. One may show that if

$$[\hat{a}_{\vec{p}},\hat{a}^\dagger_{\vec{q}}] = (2\pi)^n 2E_{\vec{p}}\delta^{(n)}(\vec{x}-\vec{y}), \tag{8}$$

then Eq. 7 does satisfy Eq. 5. Notice that the required commutation relation for the raising and lowering operators Eq. 8 is consistent with the dimensional analysis requirement that $[\hat{a}_{\vec{p}}] = [\hat{a}^\dagger_{\vec{p}}] = E^{(1-n)/2}$.

Now, from a direct application of Noether's theorem,

$$\begin{aligned}T^{\mu\nu} &= \frac{\partial\mathcal{L}}{\partial(\partial_\mu\phi)}\partial^\nu\phi - \eta^{\mu\nu}\mathcal{L}\\ &= \partial^\mu\phi\partial^\nu\phi - \eta^{\mu\nu}\left[\frac{1}{2}\partial_\alpha\phi\partial^\alpha\phi - \frac{1}{2}m^2\phi^2\right].\end{aligned} \tag{9}$$

Notice that the above $T^{\mu\nu}$ has the right dimensions for an energy momentum tensor density. In particular, $E = \int d^nxT^{00}$. Since $[\int d^nx] = E^{-n}$, we have that $[T^{\mu\nu}] = E^{n+1}$, which is consistent with, e.g., $[\partial^\mu\phi\partial^\nu\phi] = [\partial^\mu][\phi][\partial^\nu][\phi] = E\,E^{(n-1)/2}E\,E^{(n-1)/2} = E^{n+1} = [m^2\phi^2]$.

It's worth noting, however, that Eq. 9 is sometimes not the best energy momentum tensor density to use. For example, Eq. 9 has non-zero trace even for a conformal (massless) scalar field theory. An improved energy momentum tensor density [6, 13] is

$$\theta^{\mu\nu} \equiv T^{\mu\nu} - \frac{1}{4}\frac{n-1}{n}(\partial^\mu\partial^\nu - \eta^{\mu\nu}\Box)\phi^2. \tag{10}$$

One may work out from the solution for ϕ and the commutation relations Eq. 8 that

$$\hat{P}^\mu = \int d^nx\,\hat{T}^{0\mu} = \int\frac{d^np}{(2\pi)^n2p^0}p^\mu\left(\hat{a}^\dagger_{\vec{p}}\hat{a}_{\vec{p}} + 2p^0(2\pi)^n\delta^{(n)}(0)\right) \to \int\frac{d^np}{(2\pi)^n2p^0}p^\mu\hat{a}^\dagger_{\vec{p}}\hat{a}_{\vec{p}}, \tag{11}$$

where in the final step we dropped the infinite vacuum energy contribution (the potentially infinite spatial infinites are 0 by symmetry). Remember that in QFT we only care about energy differences, not the absolute energy.

We immediately see that we've fully solved our problem: our states are specified by $|\ldots n_{\vec{p}}, n_{\vec{q}},\ldots\rangle \propto \cdots(\hat{a}^\dagger_{\vec{p}})^{n_{\vec{p}}}(\hat{a}^\dagger_{\vec{q}})^{n_{\vec{q}}}\cdots|0\rangle$, where $|0\rangle$ is the ground state (i.e. the vacuum) with $\hat{a}_{\vec{p}}|0\rangle = 0$, and each state has energy $\hat{H}|\ldots n_{\vec{p}}, n_{\vec{q}},\ldots\rangle = (\ldots + n_{\vec{p}}p^0 + n_{\vec{q}}q^0 + \ldots)|\ldots n_{\vec{p}}, n_{\vec{q}},\ldots\rangle$.

We may now interpret our result: the quanta that are created and destroyed by $\hat{a}^\dagger_{\vec{p}}$ and $\hat{a}_{\vec{p}}$ have all the properties we expect of particles (energy, momentum, spin [not shown], come in discrete numbers), so we'll make the intellectual leap to interpret these quanta as particles.

In particular, let's examine

$$\phi(x)|0\rangle = \int\frac{d^np}{(2\pi)^n2p^0_{\vec{p}}}\left(\hat{a}_{\vec{p}}e^{-ip\cdot x} + \hat{a}^\dagger_{\vec{p}}e^{ip\cdot x}\right)|0\rangle = \int\frac{d^np}{(2\pi)^n2p^0}e^{ip\cdot x}|\vec{p}\rangle, \tag{12}$$

where we've defined $|\vec{p}\rangle \equiv \hat{a}^\dagger_{\vec{p}}|0\rangle$ as the usual momentum eigenstate. (More on $|\vec{p}\rangle$ in a moment.)

Eq. 12 should look very familiar. Recall that in NRQM, $|\vec{x}\rangle = \int \frac{d^np}{(2\pi)^n} e^{-i\vec{p}\cdot\vec{x}}|\vec{p}\rangle$. Eq. 12 is clearly the relativistic analog: the integration measure is changed to be relativistically invariant and the $E_{\vec{p}}$ giving the time dependence (in the Heisenberg picture) is now relativistically correct.

We will therefore interpret $\phi(x)|0\rangle$ as creating a single particle at position x^μ. Looking towards the future, when we have an interacting theory, we expect the only change to be that $\phi_0(x)$ may not create exactly one particle at x^μ (or, equivalently, may not create a particle with probability 1); rather, we may need to renormalize the weight of the wavefunctions.

Notice that our momentum eigenstates contain a funny dimension. We'll take the vacuum to be properly normalized, $\langle 0|0\rangle = 1$; i.e. $[|0\rangle] = 1$, the vacuum bra and ket are dimensionless. Since $[\hat{a}^\dagger_{\vec{p}}] = E^{(1-n)/2}$, we have that $[|\vec{p}\rangle] = E^{(1-n)/2}$.

Note also that the inner product of two momentum eigenstates is $\langle \vec{k}|\vec{p}\rangle = \langle 0|\hat{a}_{\vec{k}}\hat{a}^\dagger_{\vec{p}}|0\rangle = (2\pi)^n 2E_{\vec{p}}\delta^{(n)}(\vec{k}-\vec{p})$. Thus $[\langle \vec{k}|\vec{p}\rangle] = E^{1-n} \Rightarrow [|\vec{p}\rangle] = E^{(1-n)/2}$, consistent with the above.

Now, if $|\vec{p}\rangle$ is really a momentum eigenstate, we should find that

$$\langle \hat{P}^\mu\rangle = \frac{\langle\psi|\hat{P}^\mu|\psi\rangle}{\langle\psi|\psi\rangle}\bigg|_{|\psi\rangle=|\vec{p}\rangle} = \frac{\langle\vec{p}|\hat{P}^\mu|\vec{p}\rangle}{\langle\vec{p}|\vec{p}\rangle} = p^\mu. \tag{13}$$

Let's check:

$$\begin{aligned}\frac{\langle\vec{p}|\hat{P}^\mu|\vec{p}\rangle}{\langle\vec{p}|\vec{p}\rangle} &= \frac{\langle 0|\hat{a}_{\vec{p}}\int d^nx\hat{T}^{0\mu}\hat{a}^\dagger_{\vec{p}}|0\rangle}{\langle\vec{p}|\vec{p}\rangle} \\ &= \frac{1}{\langle\vec{p}|\vec{p}\rangle}\langle 0|\hat{a}_{\vec{p}}\int\frac{d^nq}{(2\pi)^n(2q^0}q^\mu\hat{a}^\dagger_{\vec{q}}\hat{a}_{\vec{q}}\hat{a}^\dagger_{\vec{p}}|0\rangle \\ &= \frac{1}{\langle\vec{p}|\vec{p}\rangle}\int\frac{d^nq}{(2\pi)^n 2q^0}q^\mu\langle 0|\hat{p}_{\vec{p}}\hat{a}^\dagger_{\vec{q}}(2\pi)^n 2p^0\delta^{(n)}(\vec{p}-\vec{q})|0\rangle \\ &= p^\mu\frac{\langle\vec{p}|\vec{p}\rangle}{\langle\vec{p}|\vec{p}\rangle} = p^\mu,\end{aligned} \tag{14}$$

where to get from the second line to the third line we used Eq. 11.

Since $\langle\vec{p}|\vec{p}\rangle = (2\pi)^n 2E_{\vec{p}}\delta^{(n)}(0)$, in the last line of the above derivation we formally divided ∞ by ∞ in order to obtain a finite result. This formal division of an infinite quantity by exactly the same infinite quantity to obtain a finite result is a common procedure in QFT.

Another way in which we could've arrived at a finite result above is to recognize that $|\vec{p}\rangle$ is an unphysical state: we haven't taken into account the quantum fuzziness (Heisenberg uncertainty) in the natural world. Physically, one would at best have access to a normalizable wavepacket of some width. We should really use a state

$$|\psi\rangle = \int\frac{d^np}{(2\pi)^n 2p^0}f(\vec{p})|\vec{p}\rangle, \tag{15}$$

where $f(\vec{p})$ is a smooth function whose normalization is given by

$$\langle\psi|\psi\rangle = \int\frac{d^npd^nq}{(2\pi)^n 2p^0(2\pi)^n 2q^0}f^*(\vec{p})f(\vec{q})\langle\vec{p}|\vec{q}\rangle = \int\frac{d^np}{(2\pi)^n 2p^0}|f(\vec{p})|^2 = 1. \tag{16}$$

Two related advantages to using Eq. 15 are that: 1) $|\psi\rangle$ is dimensionless, as we are used to; and 2) we don't need to divide by $\langle\psi|\psi\rangle$ (or, rather, dividing by $\langle\psi|\psi\rangle$ is trivial as it's 1).

With our $|\psi\rangle$ in hand, we may evaluate

$$\begin{aligned}
&\langle \hat{P}^\mu \rangle \Big|_{|\psi\rangle} \\
&= \langle \psi \hat{P}^\mu |\psi\rangle \\
&= \int \frac{d^n p}{(2\pi)^n 2p^0} f^*(\vec{p}) \langle 0|\hat{a}_{\vec{p}} \int \frac{d^n q}{(2\pi)^n 2q^0} q^\mu \hat{a}^\dagger_{\vec{q}} \hat{a}_{\vec{q}} \int \frac{d^n s}{(2\pi)^n 2s^0} f(\vec{s}) \hat{a}^\dagger_{\vec{s}} |0\rangle \\
&= \int \frac{d^n p d^n q d^n s}{(2\pi)^n 2p^0 (2\pi)^n 2q^0 (2\pi)^n 2s^0} f^*(\vec{p}) f(\vec{s}) q^\mu \langle 0|\hat{a}_{\vec{p}} \hat{a}^\dagger_{\vec{q}} (2\pi)^n 2q^0 \delta^{(n)}(\vec{q}-\vec{s})|0\rangle \\
&= \int \frac{d^n p}{(2\pi)^n 2p^0} |f(\vec{p})|^2 p^\mu \qquad (17)
\end{aligned}$$

Now Eq. 16 tells us that $\frac{1}{(2\pi)^n 2p^0}|f(\vec{p}|^2$ is a probability distribution for the momentum of the particle. Thus Eq. 17 is *precisely* the average momentum of the particle, which is then the average value of the momentum in the full field. Thus $\langle \hat{P}^\mu \rangle = \langle p^\mu \rangle$. For a momentum distribution whose average value is some momentum p_0^μ we have $\langle \hat{P}^\mu \rangle = p_0^\mu$. In a similar vein, we may make a very Peskin-like argument that, should $f(\vec{p})$ be highly peaked (i.e. we are experimentally examining an ensemble average of particles of a nearly definite momentum p_0^μ), like at the LHC, then we may replace $\int \ldots |f(\vec{p})|^2 p^\mu \to p_0^\mu$. Then $\langle \hat{P}^\mu \rangle = p_0^\mu$.

4. Acknowledgments

The author wishes to thank the National Research Foundation and the SA-CERN collaboration for generous support.

References

[1] A. Einstein, B. Podolsky, and N. Rosen, *Phys. Rev.*, vol. 47, pp. 777–780, 1935.
[2] B. Zwiebach, *A first course in string theory.* Cambridge University Press, 2006.
[3] J. Polchinski, *String theory. Vol. 1: An introduction to the bosonic string.* Cambridge University Press, 2007.
[4] C. Itzykson and J. B. Zuber, *Quantum Field Theory.* International Series In Pure and Applied Physics, New York: McGraw-Hill, 1980.
[5] G. F. Sterman, *An Introduction to quantum field theory.* Cambridge University Press, 1993.
[6] L. S. Brown, *Quantum Field Theory.* 1994.
[7] M. E. Peskin and D. V. Schroeder, *An Introduction to quantum field theory.* 1995.
[8] S. Weinberg, *The Quantum theory of fields. Vol. 1: Foundations.* Cambridge University Press, 2005.
[9] A. Zee, *Quantum field theory in a nutshell.* 2003.
[10] M. Srednicki, *Quantum field theory.* Cambridge University Press, 2007.
[11] S. Coleman, "Notes from Sidney Coleman's Physics 253a: Quantum Field Theory," 2011.
[12] M. D. Schwartz, *Quantum Field Theory and the Standard Model.* Cambridge University Press, 2014.
[13] C. G. Callan, Jr., S. R. Coleman, and R. Jackiw, "A New improved energy - momentum tensor," *Annals Phys.*, vol. 59, pp. 42–73, 1970.

IOP Conf. Series: Journal of Physics: Conf. Series **802** (2017) 012004 doi:10.1088/1742-6596/802/1/012004

Next-to-leading order corrections to the elastic scattering of an electron off of a static scattering center

Abdullah Khalil and W. A. Horowitz

Department of Physics, University of Cape Town, Private Bag X3, Rondebosch 7701, South Africa

E-mail: abdullah@aims.ac.za and wahorowitz@gmail.com

Abstract. We calculate the elastic scattering cross section for an electron off of a classical point source in weak-coupling perturbative quantum electrodynamics at next-to-leading order accuracy in the $\overline{\text{MS}}$ renormalization scheme. Since we use the $\overline{\text{MS}}$ renormalization scheme, our result is valid up to arbitrary large momentum transfers between the source and the scattered electron.

1. Introduction

Quantum Chromodynamics (QCD) is the accepted theory for describing the interaction between quarks, antiquarks and gluons [1]. One of the fundamental predictions of QCD is the existence of a state of matter called the Quark Gluon Plasma (QGP) [2,3]. The QGP is predicted to be formed at very high energy densities, and temperatures exceeding the energy density inside the atomic nuclei by an order of magnitude $1 - 10\,\text{GeV/ fm}^3$ and at a temperatures of order (~ 180) MeV.

One of the methods to study QCD is on the lattice [4,5]. A significant result of this theory is the temperature dependence of the energy density in QCD. Which shows a rapid change in the value of the energy density at the critical temperature T_c. This rapid change can be interpreted as a change in the number of degrees of freedom in the system. Well below T_c, there are three hadronic degrees of freedom due to the three lightest hadrons: π^+, π^- and π^0. Well above T_c, there are $2(N_c^2 - 1) + 2 \times 2 \times N_c \times N_f$ degrees of freedom from the fundamental gluons and quarks of the theory.

The high $p_\perp$ data from RHIC and LHC have been interpreted as evidence that jet quenching is due to final state energy loss, which is qualitatively well described by leading order pQCD methods [6,7]. We wish to check the self-consistency of these pQCD results, and to make the pQCD calculation more quantitative. To accomplish these two goals, we must compute the next-to-leading order contribution to the energy loss of partons in a QGP. As a first step towards this NLO pQCD calculation, we compute in this paper the NLO corrections to the elastic scattering of an electron off of a static source.

2. The leading order of the differential cross section

We consider the Lagrangian describing an electron interacting with a classical current

$$\mathcal{L} = -\frac{1}{4}F^{\mu\nu}F_{\mu\nu} + \bar{\psi}\left(i\partial\!\!\!/ - m\right)\psi - e\bar{\psi}\gamma^{\mu}\psi A_{\mu} + eJ_{\mu}A^{\mu}\,, \text{with } J^{\mu} = V^{\mu}\delta^{(4)}(\vec{x} - \vec{V}x^0)\,, \quad (1)$$

where $V^{\mu} = \left(1, \vec{0}\right)^{\mu}$ is the four velocity vector. We can easily write down the Feynman rules for this Lagrangian which will be exactly the same rules for the normal QED Lagrangian [1] in addition to that for each external source we write $-ieV^{\mu}$.

Let the momenta of the incoming electron to be p and for the outgoing electron to be p', then the momentum transfer would be $q = p' - p$. Also when we define the cross section we always have the delta function $\delta(E_{p'} - E_p)$ which requires $E_{p'} = E_p = E$. Now we write down the amplitude of the leading term using the Feynman rules

$$i\mathcal{M}_0 =$$

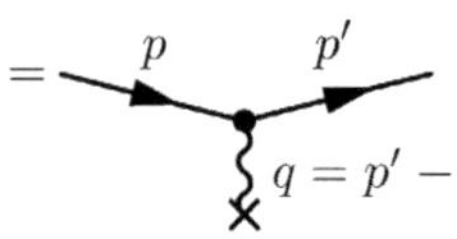

$$= \frac{i\,e^2}{q^2}\bar{u}^{s'}(p')\,\gamma^0\,u^s(p)\,. \quad (2)$$

We recall the identity $\sum_s u^s(p)\,\bar{u}^s(p) = p\!\!\!/ + m$, the trace technology and the properties of the γ-matrices. The leading term of the cross section will be

$$\left(\frac{d\sigma}{d\Omega}\right)_0 = \frac{e^4}{8\pi^2 q^4}\left(2E^2 - p'\cdot p + m^2\right)\,. \quad (3)$$

3. Renormalization of the Lagrangian

Calculating the NLO correction to the cross section require including diagrams which contain either fermion or photon loops. These loops require integrations over the loop momentum which usually diverges in 4-dimensions. In order to remove the divergences from our calculations we first should define the divergent parts by using the dimensional regularization to regularize the UV divergences and the mass regularization for the IR divergences. Then we need to renormalize our Lagrangian using the systematic way of renormalization, at the end we apply the renormalization scheme to get ride of the UV divergences.

Applying dimensional regularization requires replacing the 4 momentum integral by the integral over the momentum in d-dimensions. Which also requires rescaling the electron charge e by the factor $\mu^{\frac{4-d}{2}}$, where μ is any mass scale to make e dimensionless again [8]. At the end of the calculations, the physics (the physical parameters) should not depend on this scale.
Now let us define the Lagrangian from equation (1) in terms of the bare parameters as follows

$$\mathcal{L}_0 = -\frac{1}{4}F_0^{\mu\nu}F_{0\mu\nu} + \bar{\psi}_0(i\partial\!\!\!/ - m_0)\psi_0 - e_0\bar{\psi}_0\gamma^{\mu}\psi_0 A_{0\mu} + e_0 J_{0\mu}A_0^{\mu}\,. \quad (4)$$

Next we renormalize the fields ψ_0 and A_0^{μ} and the parameters e_0, m_0 and $J_{0\mu}$ such that

$$\psi_0 = Z_{\psi}^{\frac{1}{2}}\,\psi\,,\; A_0^{\mu} = Z_A^{\frac{1}{2}}\,A^{\mu}\,,\; Z_{\psi}\,m_0 = Z_m\,m\,,\; e_0\,Z_{\psi}Z_A^{\frac{1}{2}} = Z_e\,e\,\mu^{\frac{4-d}{2}}\,,\; \frac{Z_e}{Z_{\psi}}\,J_0^{\mu} = Z_J\,J\,. \quad (5)$$

Next we expand the renormalization parameters around 1 in terms of the counter terms (δ_{ψ}, δ_A, δ_e, δ_m and δ_J. The Lagrangian becomes

$$\begin{aligned}\mathcal{L} = &-\frac{1}{4}F_{\mu\nu}F^{\mu\nu} + \bar{\psi}(i\partial\!\!\!/ - m)\psi - e\bar{\psi}\gamma^{\mu}\psi A_{\mu} + eJ_{\mu}A^{\mu} \\ &-\frac{1}{4}\delta_A F_{\mu\nu}F^{\mu\nu} + \bar{\psi}(i\delta_{\psi}\partial\!\!\!/ - m\delta_m)\psi - e\delta_e\bar{\psi}\gamma^{\mu}\psi A_{\mu} + e\delta_j J_{\mu}A^{\mu}\,. \quad (6)\end{aligned}$$

IOP Conf. Series: Journal of Physics: Conf. Series **802** (2017) 012004 doi:10.1088/1742-6596/802/1/012004

After renormalizing our theory we would be able to write the Feynman rules for the renormalized Lagrangian which will be exactly the same rules of the renormalized QED [1].

4. NLO Corrections

In this section we compute the amplitudes due to the NLO diagrams which are the vertex correction, vacuum polarization, bremsstrahlung correction, self energy correction and the box diagram. We choose $\overline{MS}$ renormalization scheme to get ride of the UV divergences, the reason of choosing $\overline{MS}$ is that we need to extend these calculations for QCD calculations where we will deal with light quarks (i.e nearly zero mass). So we will calculate the cross section with massive electron and using $\overline{MS}$ allows us to set $m_e = 0$ safely giving a finite form of the cross section [9].

4.1. Vacuum Polarization

The amplitude for the vacuum polarization is given by

$$i\mathcal{M}_p =$$

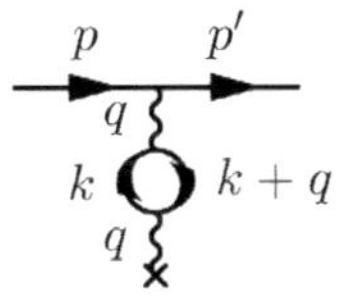

$$= \frac{-8i\alpha^2}{q^2}\,\xi(q^2)\,\bar{u}^{s'}(p')\gamma^0 u^s(p)\,, \tag{7}$$

where

$$\xi(q^2) \approx \frac{5}{18} + \frac{1}{6}\log\left(\frac{\mu^2}{-q^2}\right) \qquad , -q^2 >> m^2\,. \tag{8}$$

Now the cross section of the interference between the leading and the vacuum polarization diagrams will be

$$\left(\frac{d\sigma}{d\Omega}\right)_{PL} \approx \left(\frac{d\sigma}{d\Omega}\right)_0 \frac{\alpha}{\pi}\left(-\frac{2}{3}\log\left(\frac{\mu^2}{-q^2}\right) - \frac{10}{9}\right). \tag{9}$$

4.2. Vertex Correction

The amplitude of the vertex correction is given by

$$i\mathcal{M}_V =$$

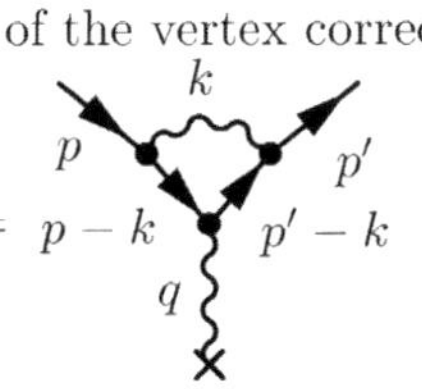

$$= \frac{4i\pi\alpha}{q^2}\bar{u}^{s'}(p')\left(\gamma^0 \cdot F_1(q^2) + \frac{i\sigma^{0\nu}q_\nu}{2m}F_2(q^2)\right)u^s(p)\,, \tag{10}$$

where $F_1(q^2)$ and $F_2(q^2)$ in the limit $-q^2 >> m^2$ will be

$$F_1(q^2) \approx \frac{\alpha}{2\pi}\left(-\frac{1}{2}\log\left(\frac{-q^2}{\mu^2}\right) - \log\left(\frac{-q^2}{m_\gamma^2}\right)\log\left(\frac{-q^2}{m^2}\right) + \frac{1}{2}\right),$$

$$F_2(q^2) \approx \frac{\alpha}{2\pi}\left(\frac{m^2}{-q^2}\log\left(\frac{-q^2}{m^2}\right)\right), \tag{11}$$

where m_γ is the photon mass to regularize the expected IR divergences. Finally the cross section of the interference between the leading and the vertex diagrams will be

$$\left(\frac{d\sigma}{d\Omega}\right)_{VL} = \left(\frac{d\sigma}{d\Omega}\right)_0 \frac{\alpha}{\pi}\left(\frac{1}{2}\log\left(\frac{\mu^2}{-q^2}\right) - \log\left(\frac{-q^2}{m_\gamma^2}\right)\log\left(\frac{-q^2}{m^2}\right) + \frac{1}{2}\right). \tag{12}$$

We note that equation (12) contains IR divergence as we send $m_\gamma = 0$ which appear as a single pole in addition to the double pole when we send both m_e and m_γ to zero.

4.3. Soft Bremsstrahlung Correction

Usually the detectors can not differentiate between the photon emitted from the vertex and the Bremsstrahlung radiation [10] which means we should add the bremsstrahlung correction. The amplitude for the soft Bremsstrahlung is given by

$i\mathcal{M}_B =$

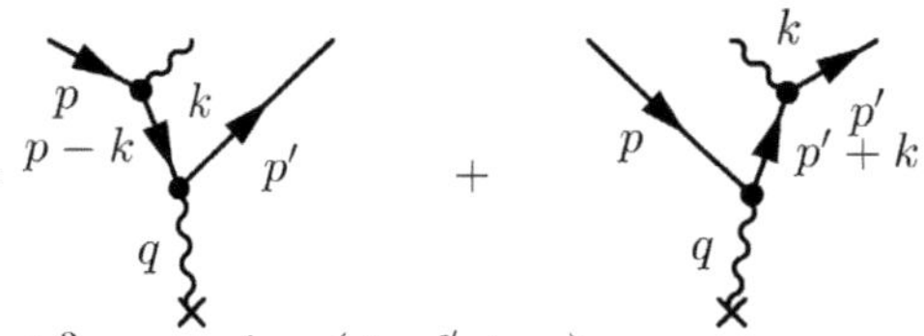

$$= \frac{ie^3}{q^2}\bar{u}^{s'}(p')\left(\gamma^0\frac{(\not{p}-\not{k}+m)}{(p-k)^2-m^2}\gamma^\alpha\varepsilon^r_\alpha(k)+\gamma^\alpha\varepsilon^r_\alpha(k)\frac{(\not{p}'+\not{k}+m)}{(p'+k)^2-m^2}\gamma^0\right)u^s(p)\,. \tag{13}$$

For the soft photon approximation $|k| << |p'-p|$. The amplitude of the soft bremsstrahlung will be given by

$$i\mathcal{M}^s_B = ie\,\mathcal{M}_0\left(\frac{p'\cdot\varepsilon^r}{p'\cdot k}-\frac{p\cdot\varepsilon^r}{p\cdot k}\right), \tag{14}$$

then the cross section due to the emission of a soft photon becomes [8, 10]

$$\left(\frac{d\sigma}{d\Omega}\right)^s_B = \left(\frac{d\sigma}{d\Omega}\right)_0\frac{\alpha}{\pi}\left(\log\left(\frac{-q^2}{m^2_\gamma}\right)\log\left(\frac{-q^2}{m^2}\right)-\log\left(\frac{-q^2}{m^2_\gamma}\right)\right). \tag{15}$$

4.4. Electron Self Energy

The amplitude of the electron self energy is given by

$-i\Sigma_2(p) =$

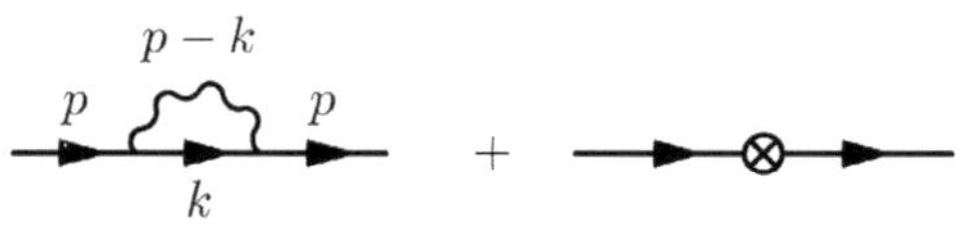

(16)

Applying the usual procedure of our renormalization scheme. The amplitude for the electron-self energy becomes

$$\Sigma_2(\not{p}) = \frac{\alpha}{4\pi}\left((\not{p}-2m)+\int_0^1 dx\,(4m-2x\not{p})\log\left(\frac{\mu^2}{(1-x)m^2-x(1-x)p^2+xm^2_\gamma}\right)\right). \tag{17}$$

The Fourier transform of the two point correlation function of the electron self energy is given by [1]

$$\begin{aligned}\int d^4x\,\langle\Omega|\,T(\psi(x)\bar{\psi}(0))\,|\Omega\rangle\,e^{ip\cdot x} &= \frac{i}{\not{p}-m}+\frac{i}{\not{p}-m}\left(\frac{\Sigma(\not{p})}{\not{p}-m}\right)+\frac{i}{\not{p}-m}\left(\frac{\Sigma(\not{p})}{\not{p}-m}\right)^2+\dots\\ &= \frac{i}{\not{p}-m-\Sigma(\not{p})}\,.\end{aligned} \tag{18}$$

This means that the pole is shifted by $\Sigma(\not{p})$, so the renormalized mass is not the physical mass and the residue of this pole is no longer one [9]. So our goal now is to find the correction to the residue and the relation between the renormalized mass m and the physical mass m_e and the correction to the residue. The physical mass can be given by the position of the pole where we have

$$m_e = m\left(1 + \frac{\alpha}{4\pi}\left(4 + 3\log\left(\frac{\mu^2}{m^2}\right)\right) + O(\alpha^2)\right) . \tag{19}$$

The inverse of the residue is given by

$$R^{-1} \approx 1 - \frac{\alpha}{4\pi}\left(2\log\left(\frac{m^2}{m_\gamma^2}\right) - \log\left(\frac{\mu^2}{m^2}\right) - 4\right) + \mathcal{O}(\alpha^2) . \tag{20}$$

We should multiply the amplitude by $R^{1/2}$ for each external leg, which means that we should multiply the differential cross section by R^2 [9]. The contribution of the self energy diagram at one loop is zero, since the contribution of the diagram due to the self energy of the incoming electron is exactly the same as the contribution due to the self energy of the outgoing electron with a relative sign difference.

4.5. Box Correction

The amplitude of the box diagram is given by

$i\mathcal{M}_{BO} =$ 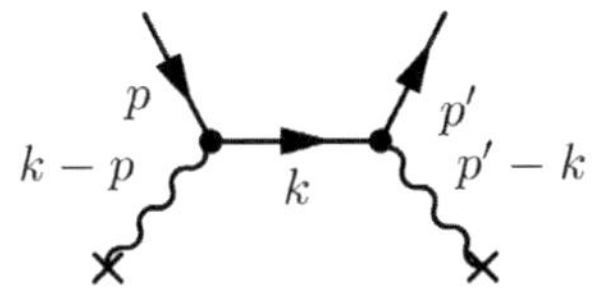 . (21)

It is obvious that the box diagram does not contain any ultraviolet divergences, so we do not need to perform dimensional regularization. We use the trick made by R. Dalitz [11] to simplify the integrals in this diagram. The differential cross section due to the interference between the leading and the box amplitudes will be

$$\left(\frac{d\sigma}{d\Omega}\right)_{BOL} = \frac{-\pi\alpha^3 E}{|q|\, q^2}\left(\frac{|q|}{p} - 1\right) . \tag{22}$$

4.6. The Differential Cross Section to the NLO Correction

We add the differential cross sections due to the leading order and all NLO corrections. But we should be careful because using $\overline{MS}$ renormalization scheme forces us to multiply the total cross section by the value of R^2 which is in $\mathcal{O}(\alpha)$ so the only affected term is the leading term as follows

$$\left(\frac{d\sigma}{d\Omega}\right) = \left(\frac{d\sigma}{d\Omega}\right)_0 \left[1 + \frac{\alpha}{\pi}\left(\frac{3}{2}\log\left(\frac{m^2}{-q^2}\right) - \frac{2}{3}\log\left(\frac{\mu^2}{-q^2}\right) - \frac{47}{18}\right)\right] + \frac{\pi\alpha^3 E}{|q|\, q^2}\left(\frac{|q|}{p} - 1\right) + \mathcal{O}(\alpha^4) \tag{23}$$

We note that equation (23) is not finite as we send $m = 0$, which arises from the emission of photon collinear with the incoming and outgoing electrons through the bremsstrahlung process. Which means we should also including the initial and final state of the hard bremsstrahlung corrections according to the KLN theorem [12].

We have mentioned that the physical parameters shouldn't depend on μ, but we see that μ appears in the final formula for the differential cross section, which means the renormalized parameters depend on μ and we find

$$\beta(\alpha) = \mu\,\frac{d\alpha}{d\mu} = \frac{2\alpha^2}{3\pi} , \qquad \gamma_m(\alpha) = \frac{\mu}{m}\frac{dm}{d\mu} = \frac{-3\alpha}{2\pi} , \tag{24}$$

then it is straightforward to check that

$$\frac{dm_e}{d\mu} = 0 \quad , \quad \frac{d}{d\mu}\left(\frac{d\sigma}{d\Omega}\right) = 0 \,. \tag{25}$$

5. Conclusion

In this project we calculated the elastic differential cross section, including the next-to-leading order corrections of an electron scattered by a classical static point charge. These corrections come from the photon self energy, vertex, soft Bremsstrahlung and the box diagrams. We found that all ultraviolet divergences due to loop integrals are absorbed by the counter terms in the $\overline{\text{MS}}$ renormalization scheme. We also saw that all infrared divergences due to the zero mass of the photon cancel, where the infrared divergence from the vertex correction cancels exactly with the one from the soft and collinear Bremsstrahlung. But we still need to add the hard bremsstrahlung correction due to the final and initial states [12] to obtain finite formula of the cross section when we take the electron mass m_e to zero ; equivalently, we expect a result that is valid up to arbitrary large momentum exchange.

The next step is to extend these calculations to derive a formula for the energy lost by high momentum particles propagating through weakly coupled plasmas in thermal field theory, including the next-to-leading-order corrections due to the emission of very high energy particles. This is very interesting for the study of the QGP and for understanding the nature of the degrees of freedom at energy densities $(1-3)T_c$ from lattice QCD.

Acknowledgments

Many thanks to SA-CERN, the South African National Research Foundation (NRF), University of Cape Town (UCT) and the African Institute for Mathematical Sciences (AIMS) for their support.

References

[1] M. E. Peskin and D. V. Schroeder, *An introduction to Quantum Field Theory.* Westview, 1995.
[2] J. C. Collins and M. J. Perry, "Superdense matter: neutrons or asymptotically free quarks?," *Physical Review Letters*, vol. 34, no. 21, p. 1353, 1975.
[3] G. Baym and S. Chin, "Can a neutron star be a giant MIT bag?," *Physics Letters B*, vol. 62, no. 2, pp. 241–244, 1976.
[4] F. Karsch, E. Laermann, and A. Peikert, "The pressure in 2, 2+ 1 and 3 flavour QCD," *Physics Letters B*, vol. 478, no. 4, pp. 447–455, 2000.
[5] F. Karsch, "Lect. notes phys. 583, 209 (2002)," *arXiv preprint hep-lat/0106019*, 2002.
[6] M. Djordjevic and M. Gyulassy, "Heavy quark radiative energy loss in qcd matter," *Nuclear Physics A*, vol. 733, no. 3, pp. 265–298, 2004.
[7] M. Gyulassy, P. Lévai, and I. Vitev, "Reaction operator approach to non-abelian energy loss," *Nuclear Physics B*, vol. 594, no. 1, pp. 371–419, 2001.
[8] C. Itzykson and J.-B. Zuber, *Quantum field theory.* Courier Corporation, 2006.
[9] M. Srednicki, *Quantum Field Theory.* Cambridge University Press, 2007.
[10] F. Mandl and G. Shaw, *Quantum Field Theory.* John Wiley & Sons, 2010.
[11] R. Dalitz, "On higher born approximations in potential scattering," *Proceedings of the Royal Society of London. Series A. Mathematical and Physical Sciences*, vol. 206, no. 1087, pp. 509–520, 1951.
[12] T.-D. Lee and M. Nauenberg, "Degenerate systems and mass singularities," *Physical Review*, vol. 133, no. 6B, p. B1549, 1964.

IOP Conf. Series: Journal of Physics: Conf. Series **802** (2017) 012005 doi:10.1088/1742-6596/802/1/012005

The evolution of gauge couplings and the Weinberg angle in 5 dimensions for an SU(3) gauge group

Mohammed Omer Khojali[a,1], **A. S. Cornell**[a,2] **and Aldo Deandrea**[b,3]

[a] National Institute for Theoretical Physics; School of Physics and Mandelstam Institute for Theoretical Physics, University of the Witwatersrand, Wits 2050, South Africa

[b] Universite de Lyon, F-69622 Lyon, France; Universite Lyon 1, CNRS/IN2P3, UMR5822 IPNL, F-69622 Villeurbanne Cedex, France

E-mail: [1]khogali11@gmail.com, [2]alan.cornell@wits.ac.za, [3]deandrea@ipnl.in2p3.fr

Abstract. We test in a simplified 5-dimensional model with SU(3) gauge symmetry, the evolution equations of the gauge couplings of a model containing bulk fields, gauge fields and one pair of fermions. In this model we assume that the fermion doublet and two singlet fields are located at fixed points of the extra-dimension compactified on an S^1/Z_2 orbifold. The gauge coupling evolution is derived at one-loop in 5-dimensions, for the gauge group $G = SU(3)$, and used to test the impact on lower energy observables, in particular the Weinberg angle. The gauge bosons and the Higgs field arise from the gauge bosons in 5 dimensions, as in a gauge-Higgs model. The model is used as a testing ground as it is not a complete and realistic model for electroweak interactions.

1. Introduction

A gauge theory defined in more than four dimensions has many attractive features, where interactions at low energies may be truely unified and some of the distinct fields in four dimensions can be integrated as a single multiplet in higher dimensions, like in gauge-Higgs models, where the Higgs fields can be a component of 5-dimensional gauge fields. Note also that the topology and structure of the extra-dimensional space provides new ways of breaking symmetries [1]. The simplest theories of this type have problems in reproducing low energy observables, such as the Weinberg angle, the SM fermion content and where Yukawa couplings are different from the gauge couplings [2].

In this paper we shall discuss the gauge coupling evolution for a model which contains a pair of left-handed and right-handed fermion matter fields located at different fixed points. Our model shall contain a heavy bulk fermion, which has quantum numbers allowing couplings to both the boundary fermions. This matter field is introduced as a representation of the unified group $G = SU(3)$, where this gauge group is broken to a subgroup H. This unified $SU(3)$ group is the gauge symmetry of the electroweak interaction and a Higgs doublet field, and it is not the gauge group of the strong interaction [3].

Note that H (the subgroup of G) is not broken by the vacuum expectation value (vev) of the scalar fields (under which the vev of the scalar fields is invariant). So we can correspondingly divide the generators of G into two sets: the unbroken ones, that annihilate the vacuum, and the broken ones, the orthogonal set [4]. According to the Goldstone theorem each broken generator in the coset G/H is associated to an independent massless scalar (Goldstone bosons), carrying the same quantum numbers as the generators [5]:

$$G/H = \frac{SU(3)}{SU(2) \times U(1)} \Rightarrow \dim(G/H) = 8 - (3+1) = 4. \tag{1}$$

In the case of the bulk fields, the standard Yukawa coupling can originate only from higher-dimensional gauge couplings, but in the case of the boundary localised matter fields the standard Yukawa coupling cannot be directly introduced [6]. The gauge bosons arise from the 4-dimensional components of the 5-dimensional gauge fields, whilst the Higgs field arises from the internal components of the gauge group $G = SU(3)$ compactified on an S^1/Z_2 orbifold; and it can be described as a circle of radius R where the opposite points can be identified by the action of the Z_2 orbifold [7]. This orbifold is given as $Z_2 : y \to -y$, so our physical space is the interval $y \in [0, \pi R]$ and it has two fixed points at $y = 0$ and $y = \pi R$. The generic 5-dimensional bulk fields, for example $A_\mu(x^\mu, y)$, have the following transformation property:

$$A_\mu(x^\mu, -y) \to PA_\mu(x^\mu, y). \tag{2}$$

It also has a $U(1)$ transformation, under the boundary condition of the fifth coordinate $y \to y + \pi R$ [8]:

$$A_\mu(x^\mu, y + \pi R) \to PA_\mu(x^\mu, y). \tag{3}$$

The $A_5(x^\mu, y)$ gauge field has the following transformation property:

$$A_5(x^\mu, y) \to PA_5(x^\mu, y)P^{-1} - iP^{-1}\partial_y P, \tag{4}$$

where our Z_2 orbifold boundary condition is given in the following way:

$$P = e^{i\vartheta^a(x^\mu, y)T^a}. \tag{5}$$

Here we are considering that the gauge transformation ϑ^a is linear in y.

Taking our gauge field to be along the third isospin direction, which means that the Z_2 orbifold boundary condition can then be written in the following way:

$$P = e^{i\pi\lambda_3} = \begin{pmatrix} -1 & 0 & 0 \\ 0 & -1 & 0 \\ 0 & 0 & 1 \end{pmatrix}, \tag{6}$$

where λ_a are the standard $SU(3)$ Gell-Mann matrices, normalised as $Tr(\lambda_a\lambda_b) = 2\delta_{ab}$. The group Z_2 acts on the tours as π rotations, the orbifold projection P breaking the gauge group G in 4-dimensions to the subgroup $H = SU(2) \times U(1)$ of the projection P. The massless 4-dimensional fields are the gauge bosons A^a_μ in the adjoint of H and the charged scalar doublet arises from the internal components $A^{\underline{a}}_5$ of the gauge field [9].

The brane fields of the model we shall focus on consist of a left-handed fermion doublet $Q_L = (u_L, d_L)$, and two right-handed fermion singlets u_R and d_R. We are going to assume that the doublet and the two singlet fields are located respectively at positions y_1 and y_2, which equals to either 0 or πR. In this model we are neglecting the bulk to boundary couplings, because we are trying to control the Yukawa coupling terms, which are present in the model. We are also trying to simplify the model.

The Lagrangian for the bulk fields, gauge fields and the pair of fermions is given by:

$$\begin{aligned}\mathcal{L}_{matter} &= \sum_a \Big[i\bar{\psi}_a(x,y)\not{D}_5\psi_a(x,y) + i\bar{\tilde{\psi}}_a(x,y)\not{D}_5\tilde{\psi}_a(x,y) + \bar{\psi}_a(x,y)M_a\tilde{\psi}_a(x,y) \\ &+\bar{\tilde{\psi}}_a(x,y)M_a\psi_a(x,y)\Big] + \delta(y-y_1)\Big[i\tilde{Q}_L(x,y)\not{D}_\mu Q_L(x,y)\Big] \\ &+\delta(y-y_2)\Big[i\bar{d}_R(x,y)\not{D}_\mu d_R(x,y) + i\bar{u}_R(x,y)\not{D}_\mu u_R(x,y)\Big], \end{aligned} \quad (7)$$

where $\not{D}_\mu$ and $\not{D}_5$ are the 4-dimensional and 5-dimensional covariant derivatives respectively, and are related by the following equality

$$\not{D}_5 = \not{D}_\mu + i\gamma_5 D_5. \quad (8)$$

$$\not{D}_M = \gamma^M\partial_M - i\gamma^M g_M A_M^a T^a, \quad (9)$$

where $M = (\ \mu$ or $5)$, the Hermitian matrix $\gamma_5 = i\gamma_\mu$, T^a are the generators of the Lie algebra of the gauge group G, A_μ^a are the 4-dimensional gauge bosons and the scalar fields A_5^a are identified with the components of the Higgs field [10].

In the fundamental representation of the gauge group G, the mode expansion for the left-handed ψ_L and the right-handed ψ_R bulk fermion is

$$\psi_{aL}(y) = \sum_{-\infty}^{\infty} \eta_n \frac{1}{\sqrt{2\pi R}} \sin\left(\frac{ny}{R}\right)\psi_{aL}^n(x), \quad (10)$$

$$\psi_{aR}(y) = \sum_{-\infty}^{\infty} \eta_n \frac{1}{\sqrt{2\pi R}} \cos\left(\frac{ny}{R}\right)\psi_{aR}^n(x). \quad (11)$$

By adding equations (10) and (11) one can get the corresponding Fourier decomposition of a generic bulk fermion

$$\psi_a(y) = \sum_{n=-\infty}^{\infty} \eta_n \frac{1}{\sqrt{2\pi R}}\left[\sin\left(\frac{ny}{R}\right)\psi_{aL}^n(x) + \cos\left(\frac{ny}{R}\right)\psi_{aR}^n(x)\right], \quad (12)$$

where the factor η_n is defined to be 1 for n = 0 and $1/\sqrt{2}$ for $n \neq 0$, which means we can rewrite the bulk fermion in equation (12) as

$$\psi_a(y) = \frac{1}{\sqrt{2\pi R}}\psi_{aR}^0(x) + \frac{1}{2\sqrt{\pi R}}\sum_{n=1}^{\infty}\left[\sin\left(\frac{ny}{R}\right)\psi_{aL}^n(x) + \cos\left(\frac{ny}{R}\right)\psi_{aR}^n(x)\right]. \quad (13)$$

The 4-dimensional Lagrangian for the bulk fermion ψ_a is written as

$$\mathcal{L}_{4D}^{\psi_a} = \int_0^{\pi R}\left[\bar{\psi}_a(y) i\not{D}_5\psi_a(y)\right]dy, \quad (14)$$

where integrating out the y coordinate one can get

$$\begin{aligned}
\mathcal{L}_{4D}^{\psi_a} &= \frac{1}{2}\Bigg[\bar{\psi}_{aR}^0(x)(\gamma^\mu\partial_\mu - \gamma_\mu\partial_5)\psi_{aR}^0(x) + i\gamma_\mu g_5 T^a \bar{\psi}_{aR}^0(x) A_5^a \psi_{aR}^0(x) \\
&\quad - i\gamma^\mu g_\mu T^a \bar{\psi}_{aR}^0(x) A_\mu^a \psi_{aR}^0(x) + \frac{1}{4}\Bigg(\sum_{n=1}^{\infty} \bar{\psi}_{aL}^n(x)(\gamma^\mu\partial_\mu - \gamma_\mu\partial_5)\psi_{aL}^n(x) \\
&\quad + i\gamma_\mu g_5 T^a \sum_{n=1}^{\infty} \bar{\psi}_{aL}^n(x) A_5^a \psi_{aL}^n(x) - i\gamma^\mu g_\mu T^a \sum_{n=1}^{\infty} \bar{\psi}_{aL}^n(x) A_\mu^a \psi_{aL}^n(x) \\
&\quad + \sum_{n=1}^{\infty} \bar{\psi}_{aR}^n(x)(\gamma^\mu\partial_\mu - \gamma_\mu\partial_5)\psi_{aR}^n(x) + i\gamma_\mu g_5 T^a \sum_{n=1}^{\infty} \bar{\psi}_{aR}^n(x) A_5^a \psi_{aR}^n(x) \\
&\quad - i\gamma^\mu g_\mu T^a \sum_{n=1}^{\infty} \bar{\psi}_{aR}^n(x) A_\mu^a \psi_{aR}^n(x)\Bigg)\Bigg]. \qquad (15)
\end{aligned}$$

We can obtain the 4-dimensional Lagrangian for the bulk fermion, $\tilde{\psi}_a$, in similar way as in the case of the bulk fermion, ψ_a, by replacing ψ_a by $\tilde{\psi}_a$ in equation (15).

Now let us move to the case of the 4-dimensional left-handed fermion doublet, where the Fourier decomposition for that field is written as

$$Q_L(y) = \frac{1}{\sqrt{2\pi R}} Q_L^0(x) + \frac{1}{2\sqrt{\pi R}} \sum_{n=1}^{\infty} \left[\cos\left(\frac{ny}{R}\right) Q_L^n(x) + \sin\left(\frac{ny}{R}\right) Q_R^n(x)\right]. \qquad (16)$$

The 4-dimensional Lagrangian for the left-handed fermion doublet is given by

$$\mathcal{L}_{4D}^{Q_L} = \int_0^{\pi R} dy \delta(y - y_1) \left[\bar{Q}_L i \not{D}_\mu Q_L\right], \qquad (17)$$

where as we mentioned before, the $\delta(y - y_1)$ is needed as the left-handed fermion doublet is located at position y_1, which is equal to either 0 or πR. By integrating out the y coordinate one can get

$$\mathcal{L}_{4D}^{Q_L} = \frac{1}{4\pi R}\left[\bar{Q}_L^0(x)[i\gamma^\mu\partial_\mu + \gamma^\mu g_\mu A_\mu^a T^a] Q_L^0(x) + \frac{1}{2}\sum_{n=1}^{\infty} \bar{Q}_L^n(x)[i\gamma^\mu\partial_\mu + \gamma^\mu g_\mu A_\mu^a T^a] Q_L^n(x)\right]. \qquad (18)$$

Finally, we can see the case of the two singlet fields which are located at position y_2, the Fourier decomposition for those fields are written as

$$d_R(y) = \frac{1}{\sqrt{2\pi R}} d_R^0(x) + \frac{1}{2\sqrt{\pi R}} \sum_{n=1}^{\infty} \left[\cos\left(\frac{ny}{R}\right) d_R^n(x) + \sin\left(\frac{ny}{R}\right) d_L^n(x)\right], \qquad (19)$$

$$u_R(y) = \frac{1}{\sqrt{2\pi R}} u_R^0(x) + \frac{1}{2\sqrt{\pi R}} \sum_{n=1}^{\infty} \left[\cos\left(\frac{ny}{R}\right) u_R^n(x) + \sin\left(\frac{ny}{R}\right) u_L^n(x)\right]. \qquad (20)$$

The 4-dimensional Lagrangian for the two singlet fields d_R and u_R is written as

$$\mathcal{L}_{4D}^{singlet} = \int_0^{\pi R} dy \delta(y - y_2) \left[\bar{d}_R i \not{D}_\mu d_R + \bar{u}_R i \not{D}_\mu u_R\right], \qquad (21)$$

where by integrating out the y coordinate one can get

$$\mathcal{L}_{4D}^{singlet} = \frac{1}{4\pi R}\Bigg[\bar{d}_R^0(x)[i\gamma^\mu\partial_\mu + \gamma^\mu g_\mu A_\mu^a T^a]d_R^0(x) + \bar{u}_R^0(x)[i\gamma^\mu\partial_\mu + \gamma^\mu g_\mu A_\mu^a T^a]u_R^0(x)$$
$$+\frac{1}{2}\sum_{n=1}^{\infty}\bar{d}_R^n(x)[i\gamma^\mu\partial_\mu + \gamma^\mu g_\mu A_\mu^a T^a]d_R^n(x) + \frac{1}{2}\sum_{n=1}^{\infty}\bar{u}_R^n(x)[i\gamma^\mu\partial_\mu + \gamma^\mu g_\mu A_\mu^a T^a]u_R^n(x)\Bigg]. \quad (22)$$

2. The gauge coupling evolution equations

Our goal is to discuss the gauge coupling evolution for the model presented in the previous section. In order to do so, we need to introduced the β-functions. This crucial object is needed to determine the evolution of the coupling constants. In general, in a theory with n-couplings g_i, we have to solve a set of coupled differential equations of the form

$$\beta_i = \mu\frac{dg_i}{d\mu} = \frac{dg_i}{dt}, \quad (23)$$

where $t = (\ln[\mu/M_Z])$. In general the β-functions depend on all the couplings and masses of the theory. We can get rid of the masses by focusing only on the universal UV relevant coefficients. For example, one can focus on the gauge coupling evolution equations, where we can write the general term for the gauge interaction of the fermions and the gauge bosons as $g\bar{\psi}\gamma^\mu\psi A_\mu$. In terms of renormalisable quantities (by rescaling)

$$\bar{\psi} = Z_{\bar{\psi}}^{1/2}\bar{\psi}^R, \quad (24)$$

$$\psi = Z_{\psi}^{1/2}\psi^R, \quad (25)$$

$$A_\mu = Z_{A_\mu}^{1/2}A_\mu^R, \quad (26)$$

where $Z_\psi^{1/2}$, $Z_{\bar{\psi}}^{1/2}$ and $Z_{A_\mu}^{1/2}$ are the renormalisation constants.
By using equations (24), (25) and (26) one can write the gauge interaction of the fermions and the gauge bosons in terms of the renormalisable quantities [11]

$$gZ_{\bar{\psi}}^{1/2}Z_{\psi}^{1/2}Z_{A_\mu}^{1/2}\bar{\psi}^R\gamma^\mu\psi^R A_\mu^R = Z_g^{1/2}g^R\bar{\psi}^R\gamma^\mu\psi^R A_\mu^R. \quad (27)$$

From the above equation one can see that

$$gZ_{\bar{\psi}}^{1/2}Z_{\psi}^{1/2}Z_{A_\mu}^{1/2} = Z_g^{1/2}g^R. \quad (28)$$

As we discussed earlier, the couplings g_i are determined by noticing that physics cannot depend on our arbitrary choice of scale μ. We have, therefore,

$$\frac{d\ln g^R}{dt} = \frac{1}{2}\frac{d\ln Z_{\bar{\psi}_R}}{dt} + \frac{1}{2}\frac{d\ln Z_{\psi_R}}{dt} + \frac{1}{2}\frac{d\ln Z_{A_\mu}}{dt} - \frac{1}{2}\frac{d\ln Z_g}{dt}. \quad (29)$$

We then need to calculate the renormalisation constants. When doing so, we usually ignore the mass terms in the propagators, since they have nothing to do with the divergent part of the one-loop diagrams. We are going to focus on the UV regime where we can neglect the m/μ dependence of β [11].

The general formula of the β-functions for the gauge couplings is given by [12]

$$16\pi^2\frac{dg_i}{dt} = b_i^{SM}g_i^3 + (b_i + S(t)\tilde{b_i})g_i^3, \tag{30}$$

where $t = \ln(S(t)/M_ZR)$, $S(t) = \mu R$ for $M_Z < \mu < \ln(1/M_ZR)$. The numerical coefficients appearing in equation (30) are given by:

$$b_i^{SM} = \left[\frac{41}{10}, -\frac{19}{6}, -7\right], \tag{31}$$

are the SM β-functions coefficients.

$$b_i = \left[\frac{10}{3}, -\frac{51}{16}, -\frac{20}{3}, \frac{3}{8}\right], \tag{32}$$

are the 0-mode β-functions coefficients.

$$\tilde{b_i} = \left[\frac{45}{16\pi}, -\frac{15}{16\pi}, -\frac{5}{\pi}, 0\right]. \tag{33}$$

are the n-mode β-functions coefficients.

3. Result and discussion

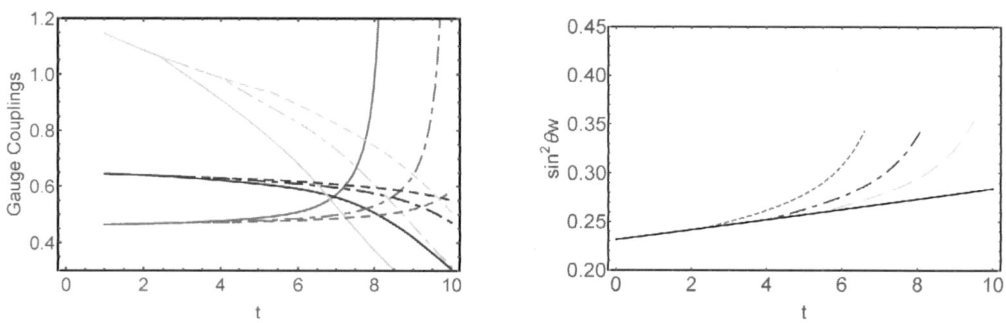

Figure 1. Left panel: Evolution of the gauge couplings g_1 (red), g_2 (blue) and g_3 (green), for three values of the $R^{-1} = 1$ TeV (solid line), 5 TeV (dot-dashed line), 20 TeV (dashed line) as a function of t. Right panel: Evolution of the Weinberg angle $\sin^2\theta_W$ with the bulk fermions, with the doublet located at position y_1 and two singlets located at position y_2, for $R^{-1} = 1$ TeV (red), $R^{-1} = 5$ TeV (blue) and $R^{-1} = 20$ TeV (green) as a function of t.

We have chosen the cut-off for our effective theory, these cut-offs are where $g_2 = g_3$, as shown in Table 1. In Figure 1, left panel, we present the evolution of the gauge couplings for the one-loop β-functions, by assuming that the bulk fermion is the top quark. We see that the three gauge couplings unify at some value of t. In the right panel we present the evolution of the Weinberg angle for the one-loop β-functions, for different values of compactification scales, for the model discussed in the previous section. When the fifth dimension contributions switch on there are large changes in the Weinberg angle up until we reach the cut-off. We can conclude that with this model the estimate of the Weinberg angle is closer to the group theoretically predicted Weinberg angle.

IOP Conf. Series: Journal of Physics: Conf. Series **802** (2017) 012005 doi:10.1088/1742-6596/802/1/012005

Scenario	t	R^{-1}	$\sin^2\theta_W$
5D $SU(3)$ gauge group	6.59	1TeV	0.335
5D $SU(3)$ gauge group	8.14	5TeV	0.345
5D $SU(3)$ gauge group	9.47	20TeV	0.348

Table 1. The cut-off and Weinberg angle for the model presented in the previous sections for the three different compactification radii $R^{-1} = 1$, 5 and 20 TeV, where $t = \ln(\mu/M_Z)$.

Acknowledgments

This work is supported by the National Research Foundation (South Africa).

References

[1] N. Haba, M. Harada, Y. Hosotani and Y. Kawamura, Nucl. Phys. B **657**, 169 (2003) [Nucl. Phys. B **669**, 381 (2003)] [hep-ph/0212035].
[2] C. Csaki, C. Grojean and H. Murayama, Phys. Rev. D **67**, 085012 (2003) [hep-ph/0210133].
[3] N. Kitazawa and Y. Sakai, Mod. Phys. Lett. A **31**, no. 07, 1650041 (2016) [arXiv:1509.03409 [hep-ph]].
[4] A. Romanino, The Standard model of particle physics, Proceedings, International Baikal Summer School, Bolshie Koty, Russia, July 23-30, 2009.
[5] Y. Nambu, Phys. Rev. 117 (1960) 648. J. Goldstone, Nuovo Cim. 19 (1961) 154. J. Goldstone, A. Salam and S. Weinberg, Phys. Rev. 127 (1962) 965.
[6] G. Burdman and Y. Nomura, Nucl. Phys. B **656**, 3 (2003) [hep-ph/0210257].
[7] I. Antoniadis, K. Benakli and M. Quiros, Phys. Lett. B **331**, 313 (1994) [hep-ph/9403290].
[8] L. J. Hall, H. Murayama and Y. Nomura, Nucl. Phys. B **645**, 85 (2002) [hep-th/0107245].
[9] C. A. Scrucca, M. Serone and L. Silvestrini, Nucl. Phys. B **669**, 128 (2003) [hep-ph/0304220].
[10] A. Masiero, C. A. Scrucca, M. Serone and L. Silvestrini, Phys. Rev. Lett. **87**, 251601 (2001) [hep-ph/0107201].
[11] M. Serone. Notes on Quantum Field Theory, Lecture notes from SISSA 2015.
[12] G. Bhattacharyya, A. Datta, S. K. Majee and A. Raychaudhuri, Nucl. Phys. B **760**, 117 (2007) [hep-ph/0608208].

Short path length pQCD corrections to energy loss in the quark gluon plasma.

Isobel Kolbe[1] **and W. A. Horowitz**[1]

[1]Department of Physics, University of Cape Town, Private Bag X3, Rondebosch 7701, South Africa

E-mail: isobel.kolbe@gmail.com

Abstract. Recent surprising discoveries of collective behaviour of low-p_T particles in pA collisions at LHC hint at the creation of a hot, fluid-like QGP medium. The seemingly conflicting measurements of non-zero particle correlations and R_{pA} that appears to be consistent with unity demand a more careful analysis of the mechanisms at work in such ostensibly minuscule systems. We study the way in which energy is dissipated in the QGP created in pA collisions by calculating, in pQCD, the short separation distance corrections to the well-known DGLV energy loss formulae that have produced excellent predictions for AA collisions. We find that, shockingly, the large formation time (compared to the $1/\mu$ Debye screening length) assumption that was used in the original DGLV calculation, results in a highly non-trivial cancellation of correction terms. We investigate the effect of relaxing the large formation time assumption in the final stages of the calculation and find that, not only is the effect of the small separation distance correction important even in large ($\sim$ 5 fm) systems, but also that the correction term dominates over the leading term at high energies.

1. Introduction

Recently, the discovery of a number of key signatures of the quark-gluon plasma (QGP) (collective behaviour [1], strangeness enhancement [2,3], and quarkonium suppression [4]) at the Relativisitc Heavy Ion Collider (RHIC) and Large Hadron Collider (LHC) have raised alarming questions regarding the presence of a QGP in small colliding systems like pp and p/dA. Jet quenching provides a means of probing the dynamics of the degrees of freedom of the QGP and is therefore also a key observable [5,6], with early experimental analysis showing signs of centrality dependent jets in p/dA collisions at RHIC and LHC [7,8].

Perturbative Quantum Chromodynamics (pQCD) has provided a basis for a number of energy loss models [9–12] that have been extremely successful, qualitatively describing the momentum dependence and angular distribution of the suppression of high-momentum, $\sim 5-150$ GeV single particle pions [13,14] and charged hadrons [15–17] from primordial hard light flavors and gluons and electrons [18–20] as well as D [21] and non-prompt J/ψ mesons [22] from open heavy flavor decays at mid rapidity in A+A systems from $\sqrt{s} = 0.2$ ATeV to 2.76 ATeV. The QGP signatures seen in p/dA collisions then call for quantitative theoretical predictions for jet tomography in small colliding systems.

However, the usual DGLV opacity expansion [23,24] derivation of the energy loss assumes a large separation distance $\Delta z \equiv z_1 - z_0 \gg \lambda_{mfp} \gg 1/\mu$ between the initial production position z_0 of the hard parent parton and the position z_1 where it scatters off a QGP medium quasiparticle.

The mean free path of the high-p_T particle is $\lambda_{mfp} = 1/\rho\sigma \sim 1 - 2$ fm while the Debye mass in an infinite, static thermal QGP of temperature $T \sim 350$ MeV is $\mu = gT \sim 0.5$ GeV, as derived from thermal field theory [25]. In pp and p/dA collisions, one expects a system of radius $\lesssim 2$ fm, suggesting that a high-p_T parton that does scatter cannot have a large separation distance between production and scattering.

In order to apply DGLV energy loss to small systems, we derive the $N = 1$ opacity generalization for all separation distances. The correction has two curious properties. The first is that the vast majority of correction terms are suppressed under the usual assumptions of eikonality, collinearity, and large gluon formation time common to all pQCD-based energy loss formalisms [26], resulting in only two diagrams with non-zero correction terms. The second is that those correction terms that do remain dominate the large separation distance DGLV expression at high (~ 100 GeV) parton energies. We will show that this second surprise is due to the breakdown of the large formation time approximation in the DGLV formalism. While the sensitivity of all energy loss formalisms to the collinear approximation has already been shown [26, 27], the present sensitivity to the large formation time approximation is both new and different from the collinear sensitivity.

2. Setup

We follow the DGLV calculation [24], treating the high-p_T eikonal parton produced at an initial point $(t_0, z_0, \mathbf{x}_0)$ inside a finite QGP, where we have used $\mathbf{p}$ to mean transverse 2D vectors, $\vec{\mathbf{p}} = (p_z, \mathbf{p})$ for 3D vectors and $p = (p^0, \vec{\mathbf{p}}) = [p^0 + p^z, p^0 - p^z, \mathbf{p}]$ for four vectors in Minkowski and light cone coordinates respectively. As in the DGLV calculation, we consider the target to be a Gyulassy-Wang Debye screened potential [28] with Fourier and color structure given by

$$\begin{aligned} V_n &= V(\vec{\mathbf{q}}_n)e^{-i\vec{q}_n\cdot\vec{x}_n} \\ &= 2\pi\delta(q^0)v(\mathbf{q}_n, q_n^z)e^{-i\vec{q}_n\cdot\vec{x}_n}T_{a_n}(R)\otimes T_{a_n}(n). \end{aligned} \tag{1}$$

The color exchanges are handled using the applicable $SU(N_c)$ generator $T_a(n)$ in the d_n dimensional representation of the target or $T_a(R)$ in the d_R dimensional representation of the high-p_T parent parton.

In light cone coordinates the momenta (defined in **??**) of the emitted gluon, the final high-p_T parton, and the exchanged medium Debye quasiparticle are

$$k = \left[xP^+, \frac{m_g^2 + \mathbf{k}^2}{xP^+}, \mathbf{k}\right], p = \left[(1-x)P^+, \frac{M^2 + \mathbf{k}^2}{(1-x)P^+}, -\mathbf{k}\right], q = [q^+, q^-, \mathbf{q}], \tag{2}$$

where the initially produced high-p_T particle of mass M has large momentum $E^+ = P^+ = 2E$ and negligible other momentum components. Notice that we include the Ter-Mikayelian plasmon effect with an effective emitted gluon mass m_g [24, 25].

A shorthand for energy ratios will prove useful notationally. Following [24], we define $\omega \approx xE^+/2 = xP^+/2$, $\omega_0 \equiv \mathbf{k}^2/2\omega$, $\omega_i \equiv (\mathbf{k} - \mathbf{q}_i)^2/2\omega$, $\omega_{(ij)} \equiv (\mathbf{k} - \mathbf{q}_i - \mathbf{q}_j)^2/2\omega$, and $\tilde{\omega}_m \equiv (m_g^2 + M^2x^2)/2\omega$.

Additionally, a number of crucial assumptions are made in line with [24]: 1) the eikonal, or high energy, approximation, for which E^+ is the largest energy scale of the problem; 2) the soft (radiation) approximation $x \ll 1$; 3) collinearity, $k^+ \gg k^-$; 4) that the impact parameter varies over a large transverse area; and, most crucially for the present work, 5) the large formation time assumption $\omega_i \ll \mu_i$, where $\mu_i^2 \equiv \mu^2 + \mathbf{q}_i^2$. These assumptions allow us to 1) (eikonal) ignore the spin of the high-p_T parton; 2) (soft) assume the source current for the parent parton varies slowly with momentum $J(p - q + k) \approx J(p + k) \approx J(p)$; 3) (collinearity) complete a separation of energy scales

$$E^+ \gg k^+ \gg k^- \equiv \omega_0 \sim \omega_{(i\dots j)} \gg \frac{(\mathbf{p} + \mathbf{k})^2}{P^+}; \tag{3}$$

and 4) take the ensemble average over the phase factors, which become $\langle e^{-i(\mathbf{q}-\mathbf{q}')\cdot\mathbf{b}}\rangle = \frac{(2\pi)^2}{A_\perp}\delta^2(\mathbf{q}-\mathbf{q}')$.

Within the above setup, we re-evaluated the 10 N=1 in opacity diagrams without the large separation distance $\Delta z \gg 1/\mu$ assumption.

3. Calculation and Results

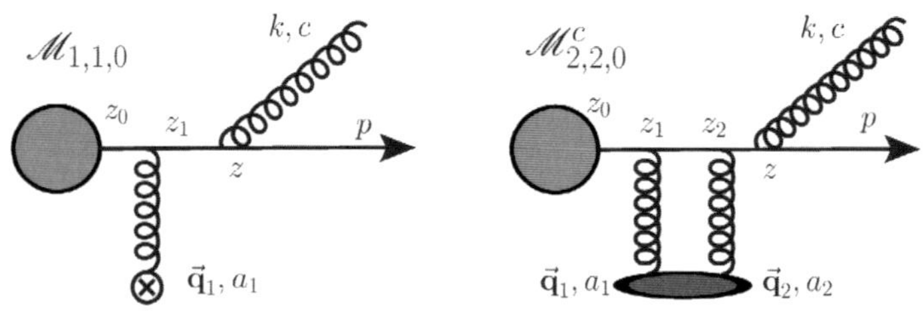

Figure 1: $\mathscr{M}_{1,1,0}$ (top panel) and $\mathscr{M}^c_{2,2,0}$ (lower panel) are the only two diagrams that have non-zero short separation distance corrections in the large formation time limit. $\mathscr{M}^c_{2,2,0}$ is the double Born contact diagram, corresponding to the second term in the Dyson series in which two gluons are exchanged with the single scattering center.

In the original evaluation of the 10 diagrams contributing to the $N = 1$ in opacity energy loss derivation, the large separation distance approximation $\Delta z \gg 1/\mu$ allowed the neglect of terms proportional to $\exp(-\mu\Delta z)$. In our reevaluation of these 10 diagrams, we retained all terms proportional to $\exp(-\mu\Delta z)$. However, we found an enormous simplification due to the large radiated gluon formation time approximation $\omega_i \ll \mu_i$: all but 2 of the 10 diagrams' 18 new small distance correction pole contributions are suppressed under the large formation time assumption. We show the two diagrams with non-zero contributions at the amplitude level $\mathscr{M}_{1,1,0}$ and $\mathscr{M}^c_{2,2,0}$ in the large formation time approximation in 1.

The full result for these two amplitudes under our approximation scheme is then

$$\begin{aligned} \mathscr{M}_{1,1,0} &\approx -J(p)e^{ipx_0}2gT_{a_1}ca_1 \int \frac{d^2\mathbf{q}_1}{(2\pi)^2} v(0,\mathbf{q}_1)e^{-i\mathbf{q}_1\cdot\mathbf{b}_1} \\ &\times \frac{\mathbf{k}\cdot\boldsymbol{\epsilon}}{\mathbf{k}^2+m_g^2+x^2M^2}\left[e^{i(\omega_0+\tilde{\omega}_m)(z_1-z_0)} - \frac{1}{2}e^{-\mu_1(z_1-z_0)}\right] \end{aligned} \tag{4}$$

$$\begin{aligned} \mathscr{M}^c_{2,2,0} &\approx J(p)e^{i(p+k)x_0} \int \frac{d^2\mathbf{q}_1}{(2\pi)^2} \int \frac{d^2\mathbf{q}_2}{(2\pi)^2} e^{-i(\mathbf{q}_1+\mathbf{q}_2)\cdot\mathbf{b}_1} \\ &\times igT_{a_2}T_{a_1}ca_2a_1 v(0,\mathbf{q}_1)v(0,\mathbf{q}_1)\frac{\mathbf{k}\cdot\boldsymbol{\epsilon}}{\mathbf{k}^2+m_g^2+x^2M^2} \\ &\times \left[e^{i(\omega_0+\tilde{\omega}_m)(z_1-z_0)} + e^{-\mu_1(z_1-z_0)}\left(1-\frac{\mu_1 e^{-\mu_2(z_1-z_0)}}{2(\mu_1+\mu_2)}\right)\right]. \end{aligned} \tag{5}$$

It should be noted that the large formation time assumption allows one to neglect terms in both the original DGLV and the current short separation distance correction. The double differential single inclusive gluon emission distribution is given by [24]

$$d^3N_g^{(1)}d^3N_J = \frac{d^3\vec{\mathbf{p}}}{(2\pi)^3 2p^0}\frac{d^3\vec{\mathbf{k}}}{(2\pi)^3 2\omega}\left(\frac{1}{d_T}\operatorname{Tr}\langle|\mathscr{M}_1|^2\rangle + \frac{2}{d_T}\Re\operatorname{Tr}\langle\mathscr{M}_0^*\mathscr{M}_2\rangle\right), \tag{6}$$

from which the energy loss, given by the energy-weighted integral over the gluon emission distribution $\Delta E = E \int dx\, x dN_g/dx$, can be calculated from the amplitudes.

Our main analytic result is then the $N = 1$ first order in opacity small distance generalization of the DGLV induced energy loss of a high-p_T parton in a QGP:

$$\Delta E^{(1)}_{ind} = \frac{C_R \alpha_s L E}{\pi \lambda_g} \int dx \int \frac{d^2\mathbf{q}_1}{\pi} \frac{\mu^2}{(\mu^2 + \mathbf{q}_1^2)^2} \int \frac{d^2\mathbf{k}}{\pi} \int d\Delta z \bar{\rho}(\Delta z) \Bigg[- \frac{2\big(1 - \cos\{(\omega_1 + \tilde{\omega}_m)\Delta z\}\big)}{(\mathbf{k} - \mathbf{q}_1)^2 + m_g^2 + x^2 M^2}$$
$$\times \left(\frac{(\mathbf{k} - \mathbf{q}_1)\cdot \mathbf{k}}{\mathbf{k}^2 + m_g^2 + x^2 M^2} - \frac{(\mathbf{k} - \mathbf{q}_1)^2}{(\mathbf{k} - \mathbf{q}_1)^2 + m_g^2 + x^2 M^2} \right)$$
$$+ \frac{1}{2} e^{-\mu_1 \Delta z} \Bigg\{ \left(\frac{\mathbf{k}}{\mathbf{k}^2 + m_g^2 + x^2 M^2} \right)^2 \left(1 - \frac{2C_R}{C_A}\right) \Big(1 - \cos\{(\omega_0 - \tilde{\omega}_m)\Delta z\}\Big)$$
$$+ \frac{\mathbf{k}\cdot(\mathbf{k} - \mathbf{q}_1)}{(\mathbf{k}^2 + m_g^2 + x^2 M^2)\big((\mathbf{k} - \mathbf{q}_1)^2 + m_g^2 + x^2 M^2\big)} \Big(\cos\{(\omega_0 - \tilde{\omega}_m)\Delta z\} - \cos\{(\omega_0 - \omega_1)\Delta z\}\Big) \Bigg\} \Bigg]. \tag{7}$$

The last four lines of 7 show the short separation distance correction term - this correction has precisely the properties one might expect: 1) for large separation distances the correction term goes to zero and 2) for vanishing separation distances the correction term also goes to zero, due to the destructive interference of the LPM effect. However, the correction also has a number of puzzling features: First is the breaking of color triviality to all orders in opacity which comes from the term proportional to $2C_R/C_A$. The second curiosity, only apparent through a numerical investigation, is that the correction term in fact dominates the original DGLV calculation at high energies.

To see this, a numerical investigation of equation (7) was performed in line with that in [24], resulting in figures 2a, 2b and 2c. The numerical analysis used the following values: $\mu = 0.5$ GeV, $\lambda_{mfp} = 1$ fm, $C_R = 4/3$, $C_A = 3$, $\alpha_s = 0.3$, $m_{charm} \equiv m_c = 1.3$ GeV and $m_{bottom} \equiv m_b = 4.75$ GeV, and the QCD analogue of the Ter-Mikayelian plasmon effect was taken into account by setting $m_{gluon} \equiv m_g = \mu/\sqrt{2}$. As in [25], kinematic upper limits were used for the momentum integrals such that $0 \le k \le 2x(1-x)E$ and $0 \le q \le \sqrt{3E\mu}$, due to finite kinematics. This choice of k_{max} guarantees that the final momentum of the parent parton is collinear to the initial momentum of the parent parton and that the momentum of the emitted gluon is collinear to the momentum of the parent parton. The fraction of momentum carried away by the radiated gluon, x was integrated over from 0 to 1. The distribution of scattering centers was assumed to be exponential in order to account for the rapidly expanding medium, $\bar{\rho}(z) = 2\exp(-2\Delta z/L)/L$.

In Fig. 2a we show the fractional energy loss of charm and bottom quarks of varying energy propagating through a 4 fm long static QGP brick. Notice first that the small distance correction term is generally an energy *gain* due to the sign of the color triviality breaking term and, second, that the size of the correction relative to the long distance DGLV result grows with energy.

In Fig. 2b we plot the fractional energy loss of charm and bottom quarks of energy $E = 10$ GeV for path lengths up to 5 fm. One sees that the small separation distance correction has a non-negligible effect even for large path lengths. Although initially unanticipated, the non-zero effect is due to the integration over all separation distances between the production point and the scattering position; even for large path lengths, some of the interaction distances between the parent parton and the target occur at separation distances that are small compared to the Debye screening scale. Additionally, the relative size of the small distance correction term and the leading DGLV result diminishes at fixed energy as the path length grows.

The main revelation of our numerical analysis is shown in Fig. 2c, which presents the fractional energy loss of 100 GeV charm and bottom quarks propagating up to 5 fm through a QGP. The

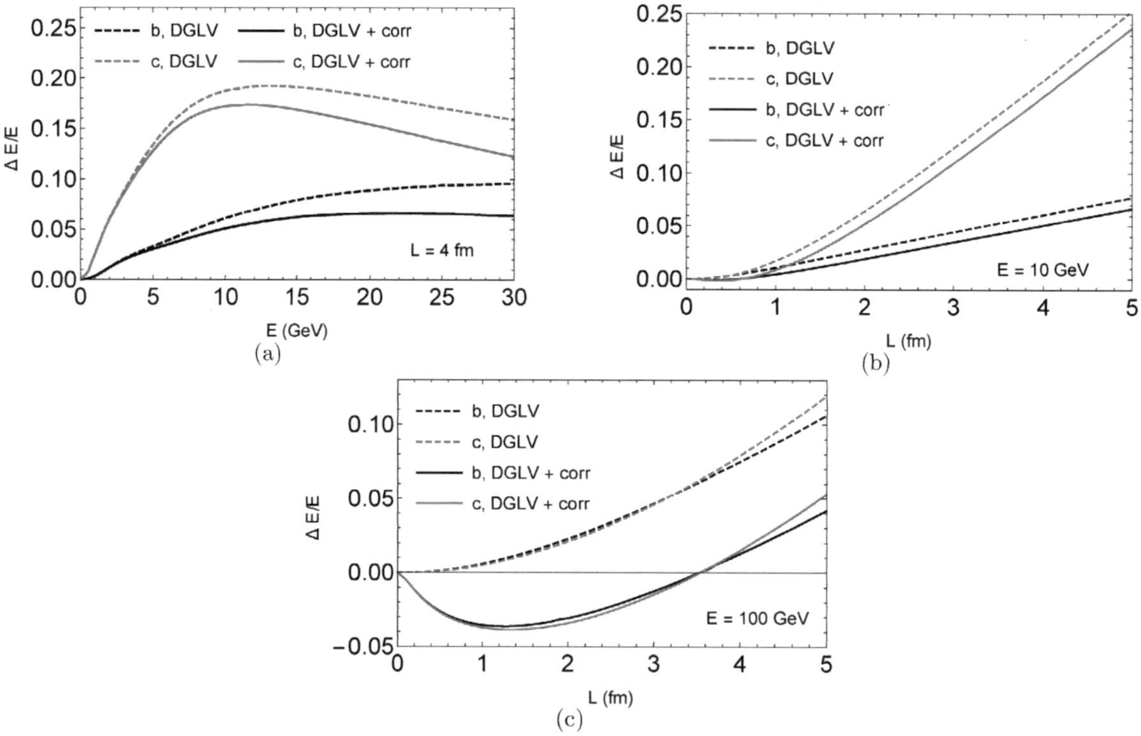

Figure 2: Fractional energy loss of charm and bottom quarks in a QGP with $\mu = 0.5$ GeV and $\lambda_{mfp} = 1$ fm for (a) fixed path length $L = 4$ fm, (b) fixed energy $E = 10$ GeV, and (c) fixed energy $E = 100$ GeV. In the figures, "DGLV" dashed curves are computed from the original $N = 1$ in opacity large separation distance DGLV formula while "DGLV + corr" solid lines are from our all separation distance generalization of the $N = 1$ DGLV result, (??).

small distance "correction" term dominates over the leading DGLV result for the first ~ 3 fm of the path.

4. Conclusions

The significant *energy gain* out to large ~ 3 fm paths seems difficult to reconcile with the measured experimental suppression of charged particles in central AA collisions at LHC [15–17]. Furthermore, the dominance of the short separation distance "correction" term at high energies, along with the universal dependence of all pQCD-based energy loss formalisms on the large formation time assumption, calls for a more detailed analysis of the validity of the large formation time assumption in pQCD-based energy loss.

5. Acknowledgments

The authors wish to thank the SA-CERN Collaboration, the South African National Research Foundation (NRF), and the National Institute for Theoretical Physics (NITheP) for their generous support. The authors also wish to thank CERN for the hospitality extended to them during the completion of part of this work. Finally, the authors wish to thank Miklos Gyulassy, Ulrich Heinz, and Carlos Salgado for valuable discussions.

References

[1] V. Khachatryan *et al.*, "Evidence for Collective Multiparticle Correlations in p-Pb Collisions," *Phys. Rev. Lett.*, vol. 115, no. 1, p. 012301, 2015.

[2] B. B. Abelev *et al.*, "Multiplicity Dependence of Pion, Kaon, Proton and Lambda Production in p-Pb Collisions at $\sqrt{s_{NN}}$ = 5.02 TeV," *Phys. Lett.*, vol. B728, pp. 25–38, 2014.

[3] C. Collaboration, "Multiplicity and rapidity dependence of strange hadrons spectra in pp, pPb, and PbPb collisions at LHC energies," 2015.

[4] J. Adam *et al.*, "Centrality dependence of inclusive J/ production in p-Pb collisions at $\sqrt{s_{\rm NN}}$ = 5.02 TeV," *JHEP*, vol. 11, p. 127, 2015.

[5] U. A. Wiedemann, "Jet Quenching in Heavy Ion Collisions," pp. 521–562, 2010. [Landolt-Bornstein23,521(2010)].

[6] A. Majumder and M. Van Leeuwen, "The Theory and Phenomenology of Perturbative QCD Based Jet Quenching," *Prog. Part. Nucl. Phys.*, vol. A66, pp. 41–92, 2011.

[7] A. Adare *et al.*, "Centrality-dependent modification of jet-production rates in deuteron-gold collisions at $\sqrt{s_{NN}}$=200 GeV," 2015.

[8] G. Aad *et al.*, "Centrality and rapidity dependence of inclusive jet production in $\sqrt{s_{\rm NN}}$ = 5.02 TeV proton-lead collisions with the ATLAS detector," *Phys. Lett.*, vol. B748, pp. 392–413, 2015.

[9] N. Armesto, M. Cacciari, A. Dainese, C. A. Salgado, and U. A. Wiedemann, "How sensitive are high-p(T) electron spectra at RHIC to heavy quark energy loss?," *Phys. Lett.*, vol. B637, pp. 362–366, 2006.

[10] W. A. Horowitz, "Heavy Quark Production and Energy Loss," *Nucl. Phys.*, vol. A904-905, pp. 186c–193c, 2013.

[11] K. M. Burke *et al.*, "Extracting the jet transport coefficient from jet quenching in high-energy heavy-ion collisions," *Phys. Rev.*, vol. C90, no. 1, p. 014909, 2014.

[12] M. Djordjevic, M. Djordjevic, and B. Blagojevic, "RHIC and LHC jet suppression in non-central collisions," *Phys. Lett.*, vol. B737, pp. 298–302, 2014.

[13] A. Adare *et al.*, "Azimuthal anisotropy of π^0 and η mesons in Au + Au collisions at $\sqrt{s_{NN}}$ = 200 GeV," *Phys. Rev.*, vol. C88, no. 6, p. 064910, 2013.

[14] B. B. Abelev *et al.*, "Neutral pion production at midrapidity in pp and Pb-Pb collisions at $\sqrt{s_{\rm NN}}$ = 2.76 TeV," *Eur. Phys. J.*, vol. C74, no. 10, p. 3108, 2014.

[15] S. Chatrchyan *et al.*, "Study of high-pT charged particle suppression in PbPb compared to *pp* collisions at $\sqrt{s_{NN}}$ = 2.76 TeV," *Eur. Phys. J.*, vol. C72, p. 1945, 2012.

[16] B. Abelev *et al.*, "Centrality Dependence of Charged Particle Production at Large Transverse Momentum in Pb–Pb Collisions at $\sqrt{s_{\rm NN}}$ = 2.76 TeV," *Phys. Lett.*, vol. B720, pp. 52–62, 2013.

[17] G. Aad *et al.*, "Measurement of charged-particle spectra in Pb+Pb collisions at $\sqrt{s_{\rm NN}}$ = 2.76 TeV with the ATLAS detector at the LHC," *JHEP*, vol. 09, p. 050, 2015.

[18] B. I. Abelev *et al.*, "Transverse momentum and centrality dependence of high-p_T non-photonic electron suppression in Au+Au collisions at $\sqrt{s_{NN}}$ = 200 GeV," *Phys. Rev. Lett.*, vol. 98, p. 192301, 2007. [Erratum: Phys. Rev. Lett.106,159902(2011)].

[19] S. Sakai, "Measurement of R_{AA} and ν_2 of electrons from heavy-flavour decays in Pb-Pb collisions at $\sqrt{s_{NN}}$ = 2.76 TeV with ALICE," *Nucl. Phys.*, vol. A904-905, pp. 661c–664c, 2013.

[20] A. Adare *et al.*, "Single electron yields from semileptonic charm and bottom hadron decays in Au+Au collisions at $\sqrt{s_{NN}}$ = 200 GeV," 2015.

[21] J. Adam *et al.*, "Transverse momentum dependence of D-meson production in Pb-Pb collisions at $\sqrt{s_{\rm NN}}$ = 2.76 TeV," 2015.

[22] S. Chatrchyan *et al.*, "Suppression of non-prompt J/ψ, prompt J/ψ, and Y(1S) in PbPb collisions at $\sqrt{s_{NN}}$ = 2.76 TeV," *JHEP*, vol. 05, p. 063, 2012.

[23] M. Gyulassy, P. Levai, and I. Vitev, "Reaction operator approach to nonAbelian energy loss," *Nucl. Phys.*, vol. B594, pp. 371–419, 2001.

[24] M. Djordjevic and M. Gyulassy, "Heavy quark radiative energy loss in QCD matter," *Nucl. Phys.*, vol. A733, pp. 265–298, 2004.

[25] S. Wicks, W. Horowitz, M. Djordjevic, and M. Gyulassy, "Elastic, inelastic, and path length fluctuations in jet tomography," *Nucl. Phys.*, vol. A784, pp. 426–442, 2007.

[26] N. Armesto *et al.*, "Comparison of Jet Quenching Formalisms for a Quark-Gluon Plasma 'Brick'," *Phys. Rev.*, vol. C86, p. 064904, 2012.

[27] W. A. Horowitz and B. A. Cole, "Systematic theoretical uncertainties in jet quenching due to gluon kinematics," *Phys. Rev.*, vol. C81, p. 024909, 2010.

[28] M. Gyulassy and X.-n. Wang, "Multiple collisions and induced gluon Bremsstrahlung in QCD," *Nucl. Phys.*, vol. B420, pp. 583–614, 1994.

The impact of additional scalar bosons at the LHC

Mukesh Kumar[a,1], Stefan von Buddenbrock[b,2], Nabarun Chakrabarty[c,3], Alan S. Cornell[a,4], Deepak Kar[b,5], Tanumoy Mandal[d,6], Bruce Mellado[b,7], Biswarup Mukhopadhyaya[c,8] and Robert G. Reed[b,9]

[a] National Institute for Theoretical Physics; School of Physics and Mandelstam Institute for Theoretical Physics, University of the Witwatersrand, Johannesburg, Wits 2050, South Africa
[b] School of Physics, University of the Witwatersrand, Johannesburg, Wits 2050, South Africa
[c] Regional Centre for Accelerator-based Particle Physics, Harish-Chandra Research Institute, Chhatnag Road, Jhusi, Allahabad - 211 019, India.
[d] Department of Physics and Astronomy, Uppsala University, Box 516, SE-751 20 Uppsala, Sweden.

E-mail: [1]mukesh.kumar@cern.ch, [2]stef.von.b@cern.ch, [3]nabarunc@hri.res.in, [4]alan.cornell@wits.ac.za, [5]deepak.kar@cern.ch, [6]tanumoy.mandal@physics.uu.se, [7]bruce.mellado@wits.ac.za, [8]biswarup@hri.res.in, [9]robert.reed@cern.ch

Abstract. In this study we consider an effective model by introducing two hypothetical real scalars, H and χ - a dark matter candidate, where the masses of these scalars are $2m_h < m_H < 2m_t$ and $m_\chi \approx m_h/2$ with m_h and m_t being the Standard Model Higgs boson and top quark masses, respectively. A distortion in the transverse momentum distributions of h in the intermediate region of the spectrum through the processes $pp \to H \to h\chi\chi$ could be observed in this model. An additional scalar, S, has been postulated to explain large $H \to h\chi\chi$ branching ratios, assuming $m_h \lesssim m_S \lesssim m_H - m_h$ and $m_S > 2m_\chi$. Furthermore, a scenario of a two Higgs doublet model (2HDM) is introduced and a detailed proposal at the present energies of the Large Hadron Collider to study the extra CP-even (h, H), CP-odd (A) and charged $(H^\pm)$ scalars has been pursued. With possible phenomenological implications, production and decay modes for these scalars are discussed. Based on the mass spectrum of H, A and $H^\pm$, the production of multi-leptons and Z+jets+missing-energy events are predicted. A specific, Type-II 2HDM model is discussed in detail.

1. Introduction

With the discovery of the scalar Higgs boson, h, the particle spectrum of the Standard Model (SM) is now complete. But there are numerous theoretical motivations to expect that the SM electroweak symmetry breaking (EWSB) mechanism is not the complete story, and the scalar spectrum responsible for it may be richer than one possessing only one neutral scalar - the Higgs boson. From this expectation, searches for new scalars, neutral or charged, are continuously being carried out both by the ATLAS and CMS collaborations in various channels. No confirmed hints along these directions has been found so far in any of these searches.

The nature of these additional particles in different theories are based on its charge, spin, mass and interactions with the SM particles. Any beyond the SM (BSM) scenario can be seen in various observables associated with the decay modes of these new particles. For example, any enhancement in the fiducial cross sections for the relevant processes could cause excesses in transverse momentum, p_T, spectrum in the intermediate regions of the decayed particles, in (pseudo)rapidity distributions or in some angular observables. Signatures of a massive object with an appropriate choice of its decay width can also be observed as an excesses in the invariant mass distribution around the mass of that object. These excesses can be understood with the help of appropriate theoretical models.

It is also noted that, if any new physics is present, one would expect that it will possibly pop-up, not only in one set of data, but in different (theoretically) correlated searches or channels with different significances. Therefore, while interpreting the excesses or anomalies, it is important to scrutinise different data sets simultaneously, which we think might be correlated in terms of simple model assumptions.

In these proceedings, we shall discuss a formalism to address some of these subjects by extending the SM with another Higgs-doublet in the theory and with an additional scalar or fermions to make the modelling appropriate and obtain the correct kinematics. Section 2 will briefly deal with the theoretical model by extending the SM with one Higgs-doublet. The phenomenological aspects, with proposed formalism, are discussed in section 3, and summary follows in section 4.

2. Theory - adding one more Higgs-doublet

We start by quoting one of the classic texts in Higgs-boson physics [1] -

"In this book we will repeatedly emphasise that many people who have studied Higgs physics do not expect a single, neutral boson of some definite mass to occur as the only direct manifestation of electroweak symmetry breaking (EWSB)."

As we all know the "minimal Higgs" in the $SU(2)_L \times U(1)_Y$ electroweak theory Higgs sector is comprised of only one complex Higgs doublet, and we have only one physical neutral Higgs scalar present in the SM spectrum, where its mass is a free parameter, not fixed by theory. In the current scenario, where this Higgs boson is almost confirmed to be $m_h \approx 125$ GeV, it is prudent to explore the implications of more complicated Higgs models in the context of extended theories. In any such theories there are basically two major constraints which need to be satisfied:

(a) First constraint comes from the experimental fact that $\rho = m_W^2/(m_Z^2 \cos^2\theta_W)$ should be very close to 1. The general formula is

$$\rho \equiv \frac{m_W^2}{m_Z^2 \cos^2\theta_W} = \frac{\sum_{T,Y}\left[4\,T\,(T+1) - Y^2\right]|V_{T,Y}|^2\, c_{T,Y}}{\sum_{T,Y} 2\,Y^2\,|V_{T,Y}|^2}, \tag{1}$$

where $\langle \Phi\,(T,Y)\rangle = V_{T,Y}$ defines the vacuum expectation values of each neutral Higgs field Φ, with T and Y specifying the total $SU(2)_L$ isospin and hypercharge of the Higgs representation to which it belongs. Here the notation, $c_{T,Y} = 1$ is for complex representation of (T,Y) and $c_{T,Y} = 1/2$ should be for real representation of $(T, Y = 0)$.

(b) Second major theoretical constraint on the Higgs sector comes from severe limits on the existence of flavour-changing neutral currents (FCNC's), which is absent in the minimal Higgs model at tree level and which must also be true in any extended models.

Keeping these issues in mind, let's extend by one more Higgs doublet and call it a two Higgs doublet model (2HDM). To date there are many versions or types of 2HDMs [2] and the key features can be summarised as follows:

(1) As an extension of the SM, this model adds new phenomena (e.g. the physical charged Higgs bosons).

(2) Since it is a minimal extension, it adds only a few new arbitrary parameters.

(3) It satisfies theoretical constraints of $\rho \approx 1$ and the absence of potentially dangerous tree-level FCNC,[1] by imposing some particular symmetries.

Considering two complex $SU(2)_L$ doublet scalar fields Φ_1 and Φ_2, the most general remormalizable scalar potential may be written as:

$$\begin{aligned} V(\Phi_1,\Phi_2) = {} & m_1^2\Phi_1^\dagger\Phi_1 + m_2^2\Phi_2^\dagger\Phi_2 - m_{12}^2\left(\Phi_1^\dagger\Phi_2 + \text{h.c.}\right) + \frac{1}{2}\lambda_1\left(\Phi_1^\dagger\Phi_1\right)^2 + \frac{1}{2}\lambda_2\left(\Phi_2^\dagger\Phi_2\right)^2 \\ & + \lambda_3\left(\Phi_1^\dagger\Phi_1\right)\left(\Phi_2^\dagger\Phi_2\right) + \lambda_4\left|\Phi_1^\dagger\Phi_2\right|^2 + \frac{1}{2}\lambda_5\left[\left(\Phi_1^\dagger\Phi_2\right)^2 + \text{h.c.}\right] \\ & + \left[\left[\lambda_6\left(\Phi_1^\dagger\Phi_1\right) + \lambda_7\left(\Phi_2^\dagger\Phi_2\right)\right]\Phi_1^\dagger\Phi_2 + \text{h.c.}\right]. \end{aligned} \tag{2}$$

After spontaneous breaking of the electroweak symmetry, five physical Higgs particles are left in the spectrum, one charged Higgs pair, $H^\pm$, one CP-odd scalar, A, and two CP-even states, H (heaviest) and h (lightest) given as:

$$H^\pm = \sin\beta\,\phi_1^\pm + \cos\beta\,\phi_2^\pm, \tag{3}$$

$$A = \sin\beta\,\mathrm{Im}\,\phi_1^0 + \cos\beta\,\mathrm{Im}\,\phi_2^0, \tag{4}$$

$$H = \cos\alpha\left(\mathrm{Re}(\phi_1^0) - v_1\right) + \sin\alpha\left(\mathrm{Re}(\phi_2^0) - v_2\right), \tag{5}$$

$$h = -\sin\alpha\left(\mathrm{Re}(\phi_1^0) - v_1\right) + \cos\alpha\left(\mathrm{Re}(\phi_2^0) - v_2\right). \tag{6}$$

Here ϕ_i^+ and ϕ_i^0 denote the $T_3 = 1/2$ and $T_3 = -1/2$ components of the ith doublet for $i = 1,2$. The angle α diagonalises the CP-even Higgs squared-mass matrix and β diagonalises both the CP-odd and charged Higgs sectors with $\tan\beta = v_2/v_1$, where $\langle\phi_i^0\rangle = v_i$ for $i = 1,2$ and $v_1^2 + v_2^2 \approx (246\,\text{GeV})^2$. This is a brief summary of how 2HDMs are constructed and further based on different choices of symmetries, couplings to quarks and leptons etc. different models can be built. Models which lead to natural flavour conservation can be named as Type-I, Type-II, Lepton-specific or Flipped 2HDMs, as detailed in Ref.[2]. In our studies we used a Type-II 2HDM for reference.

3. Phenomenology and formalism

In the previous section we discussed an extension of the Higgs boson sector, where apart from the SM Higgs boson we have extra particle spectrum of phenomenological interest. Here we describe an effective Lagrangian formalism and then follow a discussion on a specific scenario of a 2HDM, with prospects to correlate these studies for phenomenological purposes.

In all numerical analyses the codes are written in `Mathematica`, `C`$^{++}$, `Python`, models are generated with `FeynRules` which are further used for simulation through `MadGraph5`. Events are then showered and hadronized using `Pythia 8.2`, plotting and analyses followed through `Rivet` and `ROOT` package. Dark matter related constraints calculated using `microOMEGAs`.

3.1. An effective Lagrangian approach

Here we consider an effective Lagrangian approach with the introduction of two hypothetical real scalars H and χ beyond SM particle spectrum, to study phenomenology associated with Higgs boson physics. The formalism considered heavy scalar boson which decays into the SM Higgs and χ, where χ is considered as a dark matter (DM) candidate - a source of missing energy

[1] If the Higgs-fermion couplings are appropriately chosen, however, based on current FCNC requirements in some specific scenarios, this also allowed at tree level [2].

(MET). The required vertices for the studies are:

$$\mathcal{V}_H = -\frac{1}{4}\,\beta_g \kappa^{\rm SM}_{hgg}\; G_{\mu\nu}G^{\mu\nu}H + \beta_V \kappa^{\rm SM}_{hVV}\; V_\mu V^\mu H, \tag{7}$$

$$\mathcal{V}_{\rm Y} = -\frac{1}{\sqrt{2}}\left[y_{ttH}\bar{t}tH + y_{bbH}\bar{b}bH\right], \tag{8}$$

$$\mathcal{V}_{\rm T} = -\frac{1}{2}\,v\left[\lambda_{Hhh}Hhh + \lambda_{h\chi\chi}h\chi\chi + \lambda_{H\chi\chi}H\chi\chi\right], \tag{9}$$

$$\mathcal{V}_{\rm Q} = -\frac{1}{2}\lambda_{Hh\chi\chi}Hh\chi\chi - \frac{1}{4}\lambda_{HHhh}HHhh - \frac{1}{4}\lambda_{hh\chi\chi}hh\chi\chi - \frac{1}{4}\lambda_{HH\chi\chi}HH\chi\chi, \tag{10}$$

where $\beta_g = y_{ttH}/y_{tth}$ is the scale factor with respect to the SM Yukawa top coupling, y_{tth} and used to tune the effective ggH coupling as β_V is used for VVH couplings. Major constraints on the parameters of the vertices come from the relic density of DM and the DM-nuclei inelastic scattering cross sections, which left a narrow choice of mass of the DM candidate $m_\chi \sim m_h/2$ with $\lambda_{h\chi\chi} \sim [0.0006 - 0.006]$. Therefore, the masses of these scalars considered to be $2m_h < m_H < 2m_t$ and $m_\chi \approx m_h/2$, where m_t is the top quark mass. In this study if we consider a process $pp \to H \to h\chi\chi$, a distortion could be predicted in the intermediate Higgs p_T spectrum due to decay of heavy H and a recoil affect against a pair of invisible particles, χ. In order to chose appropriate values of associated couplings one must consider the constraints from all potential sources and different processes. It is to be noted, while considering various processes where the effective $Hh\chi\chi$ coupling plays a vital role one may assume this coupling to be large or small in respect of branching ratios for $H \to h\chi\chi$.

However in this effective field theory approach, we are not sure about the origin of the $Hh\chi\chi$ coupling. And in case if it is large, we can assume that this term can be generated due to the participation of a real scalar particle, S, in the intermediate state through an effective HhS coupling followed by the decay of $S \to \chi\chi$. Therefore, the inclusion of S can open up various new possibilities. In addition to the above studies, if we look over the double Higgs production modes in different decay channels like $\gamma\gamma b\bar{b}$ or $b\bar{b}b\bar{b}$ with jets etc., the vertices defined above (7)-(10) will be modified appropriately with S as an intermediate scalar and not a DM candidate,[2] with $m_h \lesssim m_S \lesssim m_H - m_h$ and $m_S > 2m_\chi$. This invokes other possibilities for the mass spectrum of S. The processes in such studies include $pp \to hS + X$ and in particular $pp \to H \to hS + X$ with respect to $pp \to H \to hh + X$, considering the available spectrum of m_S and the associated coupling parameters. There is a possibility to introduce a HSS vertex in the study, which participates further in a $H \to SS$ decay channel, as $H \to hh$ modes, with the participation of this vertex in other possible processes too. Furthermore, all other possible decay modes of S, with S into jets are possible.

3.2. Phenomenology

Considering these analyses, now we can think of H as a heavy scalar[3] of the 2HDM, and furthermore, our motive should then be to fit all the parameters like $\tan\beta$, α and the masses of A and $H^\pm$ in this specific model. But it is also a question as to whether we think of a generalised version of a 2HDM or any particular type of this model, as described in detail in Ref. [2]. On the other hand, we also need to consider experimental data on searches which will directly or in-directly affect the processes taken into consideration in this model and the signatures could be:

[2] S is a scalar particle with various decay modes and having all possible branchings to other particles, since this is no more longer stable like a DM candidate. As a result, the symmetry requirements for a gauge invariant set of vertices in the Lagrangian is different.

[3] It is to be noted that in the studies of subsection 3.1 through the effective Lagrangian, the scalar H is not necessarily assumed to be a 2HDM heavy scalar.

(1) The resonance searches gives an excess around a particular narrow mass range (with appropriate decay width approximation) in the invariant mass spectrum in di-jets or di-boson, VV, (where $V = \gamma, Z, W^{\pm}$) final states, and that provides a hint for a beyond SM particle. The masses for these resonances, m_Φ, (where $\Phi = H, A, H^{\pm}$ in 2HDMs) might be of the order of $2m_h < m_\Phi < 2m_t$ (which we considered in our previous studies for $m_\Phi = m_H$) or beyond this order like $2m_t \ll m_\Phi < \mathcal{O}(1\text{ TeV})$ or even $m_\Phi \gg \mathcal{O}(1\text{ TeV})$.

(2) In the 2HDM spectrum, we also have the possibility of charged Higgs ($H^{\pm}$) that can be produced at the LHC and these searches follows considering production cross sections and branching ratios in different decay channels. The prominent modes of decay of $H^{\pm}$ are $H^{\pm} \to tb, W^{\pm}h$ when $m_{H^{\pm}} > m_t$. Since, we consider $2m_h < m_H < 2m_t$, the decay mode of $H^{\pm} \to W^{\pm}H$ could be a prominent channel too in case of $m_{H^{\pm}} \gg m_H$.

In purview of phenomenological aspects for a CP-odd scalar, A, in 2HDMs, following salient features could be observed:

(1) In 2HDMs masses of A and $H^{\pm}$ are correlated. So if we wish to have a 2HDM with a particular mass m_A, comparably $m_{H^{\pm}}$ should also be considered. And with a known value of m_H ($2m_h < m_H < 2m_t$) and $m_h = 125$ GeV one should tune the parameters α and β accordingly.

(2) In case of a production mode of A, through the ggA vertex, there will be a need to have a scale factor β_g^A as taken in the case of H production, β_g, and considering the decay modes $A \to \gamma\gamma$ in particular needs another scale factor β_γ^A. In this respect, for consistency, one needs to control the $H \to \gamma\gamma$ decay rates via another parameter β_γ, since the form factors appearing in the calculation of $gg \to H, A$ and $H, A \to \gamma\gamma$ have a different structure and also are dependent on the masses of particles under consideration [1]. One should also study other possible decay modes A in $W^{\pm}$ or Z which is available only at loop level in 2HDMs, since W^+W^-A and AZZ couplings are absent due to CP conservation issues.

(3) This model can predict anomalously large Z+jets+MET events, we need to think of the contribution of the decay mode of $A \to ZH$, where the process should be $pp \to A \to ZH, H \to h\chi\chi$ (χ should be considered as missing energy, similar to what we considered before in $pp \to H \to h\chi\chi$). To be noted, here we need $m_A > m_Z + m_H$.

(4) In respect of point (3) mentioned above, we also consider different processes with final states with multi-lepton, tri-lepton and di-lepton channels by same-sign and opposite-sign leptons selection with jets in our phenomenological interests due to the charged Higgs, $H^{\pm}$, in 2HDMs.

(5) Since Yukawa couplings for top-quarks, y_{tth}, are well known, one should adjust the parameters α and β in such a way so that $y_{At\bar{t}}$ and $y_{Ht\bar{t}}$ must follow appropriate branchings into $A \to t\bar{t}$ and $H \to t\bar{t}$. Here we should also notice that, since y_{tth} is close to unity due to large top-quark mass, it also adds an insight for the scale of any new physics.

The phenomenology of $H^{\pm}$ itself is a subject of great detail, since one should consider either $m_{H^{\pm}} < m_t$ or $m_{H^{\pm}} > m_t$. And due to this fact the decay modes for the studies are dependent on $m_{H^{\pm}}$ following the appropriate mixing parameters α and β in 2HDMs. We consider the case in which $m_{H^{\pm}} > m_t$. The production of $H^{\pm}$ at the LHC would follow two mechanisms which have sizeable production cross sections (1) $2 \to 2$, $pp \to gb(g\bar{b}) \to tH^-(\bar{t}H^+)$ and (2) $2 \to 3$, $pp \to gg/qq' \to tH^-\bar{b} + \bar{t}H^+b$. Also $H^{\pm}$ production at hadron colliders can be studied through Drell-Yan processes for pair production, $qq \to H^+H^-$, the associated production with W bosons, $qq \to H^{\pm}W^{\pm}$, and pair production through the gluon-gluon fusion $gg \to H^+H^-$. The prominent decay modes for $H^{\pm}$ are $H^{\pm} \to tb$, $H^{\pm} \to \tau\nu$ and $H^{\pm} \to W^{\pm}h$. Here, with the allowed vertices in the 2HDM, one could think of channels where $H^{\pm}$ couples with H and further $H \to h\chi\chi$ gives opportunities to study this mode in terms of χ or missing energy aspects. The decay mode

$H^{\pm} \to W^{\pm}H$ as discussed before can be included in the studies as a prominent channel. The phenomenology of $H^{\pm}$ also depends on whether (i) $m_h < m_H < m_A$ or (ii) $m_h < m_A < m_H$, since $m_{H^{\pm}}$ could be consider as heavy as m_A.

As noticed, in all our studies we are considering an extra scalar χ as a dark matter candidate - as a signature of missing energy. But while considering various other excesses in data it may not be suitable to consider χ as a scalar. So we need to characterise χ in other possible theories which could give appropriate reasons to consider the scale factors β_i's introduced in the previous discussion, since the production (decay) modes for H/A are through gg ($\gamma\gamma$) loops and there χ may be introduced in loops. Suppose the events with $\gamma\gamma$ excess scenarios, one may introduce an extra particle in the studies, with possible ranges of m_A, and then A could be produced or decay via loops of massive coloured fermions. These extra particles could not be a fourth generation of the SM particles. Simple possibilities for these extra fermions maybe: (i) a single vector-like charges 2/3 quark. (ii) an isospin doublet of vector-like charge 2/3 and -1/3 quarks, (iii) an isodoublet and two singlet charge 2/3 and -1/3 quarks, and (iv) a complete vector-like generation including leptons as well as quarks. In this respect we should consider all four characteristics of χ, as vector-like fermions (VLF). Similar studies can follow for $W^{\pm}$ and Z decay modes of A.

3.3. A Two-Higgs doublet model

Now by knowing the requirements discussed above, we need to investigate a particular scenario of a 2HDM to accommodate and constrain the parameter space of this model. To start with the production mechanism of the scalars h, H and a pseudo scalar A, we should notice the fact that at the LHC the dominant mode of production is through gg-fusion. However, $gg\phi$, ($\phi = h, H, A$) tree level couplings are not available in 2HDMs was the case for the SM Higgs also, and is only possible through various quark-loop contributions. The generalised functional form is given as (see, e.g. Ref. [1])

$$\tau\left[1 + (1-\tau) f(\tau)\right] \to \tau\left[\xi^{\phi} + \left(1 - \tau\xi^{\phi}\right) f(\tau)\right], \tag{11}$$

where $\xi^{\phi} = 1$ for $\phi = h, H$ and $\xi^{\phi} = 0$ for $\phi = A$. τ is the ratio of masses of particles involved in the one-loop calculation. A similar functional form is true for $\gamma\gamma\phi$ vertices with a multiplication by a factor of -2 in Eq. (11), with appropriate $f(\tau)$. A proper calculation demands involved mixing parameters and couplings in that particular type of 2HDM model. So effectively depending on a choice of τ limits, $f(\tau)$ is correspondingly chosen and therefore we can incorporate those factors in a coupling parameter, $\kappa_{gg\phi}$ and $\kappa_{\gamma\gamma\phi}$. Accordingly we can write the Lagrangian for these interactions as,

$$\mathcal{L}_{gg\phi} = -\frac{1}{4}\kappa_{ggh} G_{\mu\nu^b} G^{\mu\nu b} h - \frac{1}{4}\kappa_{ggH} G^b_{\mu\nu} G^{\mu\nu b} H - \frac{1}{4}\kappa_{ggA} G^b_{\mu\nu} \tilde{G}^{\mu\nu b} A, \tag{12}$$

$$\mathcal{L}_{\gamma\gamma\phi} = -\frac{1}{4}\kappa_{\gamma\gamma h} F_{\mu\nu} F^{\mu\nu} h - \frac{1}{4}\kappa_{\gamma\gamma H} F_{\mu\nu} F^{\mu\nu} H - \frac{1}{4}\kappa_{\gamma\gamma A} F_{\mu\nu} \tilde{F}^{\mu\nu} A. \tag{13}$$

In our choice of a 2HDM, $\kappa_{gg\phi}$ and $\kappa_{\gamma\gamma\phi}$ are give as

$$\kappa_{ggh} = \frac{\alpha_s}{3\pi v}\frac{3}{4}\mathcal{F}_{12}\left[\left(\frac{m_h}{2m_t}\right)^2\right], \quad \kappa_{\gamma\gamma h} = \frac{e^2}{4\pi^2 v}\frac{47}{18}\mathcal{S}_{12}\left[\left(\frac{m_h}{2m_W}\right)^2, \left(\frac{m_h}{2m_t}\right)^2\right], \tag{14}$$

$$\kappa_{ggH} = \beta^H_g \frac{\alpha_s}{3\pi v}, \quad \kappa_{\gamma\gamma H} = \beta^H_\gamma \frac{e^2}{4\pi^2 v}, \quad \text{and } \kappa_{ggA} = \beta^A_g \frac{\alpha_s}{3\pi v}, \quad \kappa_{\gamma\gamma A} = \beta^A_\gamma \frac{e^2}{4\pi^2 v}. \tag{15}$$

Functions $\mathcal{F}_{12}$ and $\mathcal{S}_{12}$ are approximately Eq. (11) with appropriate $f(\tau)$ in a proper limit of τ (e.g. see [1]). The multiplicative factors $\beta^{H,A}_{g(\gamma)}$ are introduced to control the rates for gg and

$\gamma\gamma$-channels. Interactions with electroweak vector bosons (V) $W^{\pm}, Z$ and the photon field, A_μ, with ϕ and $H^{\pm}$, are given as

$$\mathcal{L}_{VV\phi} = \frac{2M_W^2}{v}\cos(\beta-\alpha)W_\mu^+W^{-\mu}H + 2\frac{M_W^2}{v}(\sin(\beta-\alpha))\,W_\mu^+W^{-\mu}h + \frac{M_Z^2}{v}\cos(\beta-\alpha)Z_\mu Z^\mu H + \frac{M_Z^2}{v}(\sin(\beta-\alpha))\,Z_\mu Z^\mu h, \tag{16}$$

and,

$$\begin{aligned}\mathcal{L}_{V\phi\phi} =& \frac{M_W}{v\cos\theta_W}\sin(\beta-\alpha)Z_\mu\left(A\partial_\mu H - H\partial_\mu A\right) + \frac{M_W}{v\cos\theta_W}\cos(\beta-\alpha)Z_\mu(A\partial_\mu h - h\partial_\mu A)\\ &+ i\frac{M_W}{v}\frac{(2\cos^2\theta_W-1)}{\cos\theta_W}Z_\mu\left(H^-\partial_\mu H^+ - H^+\partial_\mu H^-\right) + ieA_\mu\left(H^-\partial_\mu H^+ - H^+\partial_\mu H^-\right)\\ &+ \left[i\frac{M_W}{v}\sin(\beta-\alpha)\left(W^{-\mu}H\partial_\mu H^+ - W^{-\mu}H^+\partial_\mu H\right) + \text{h.c}\right]\\ &+ \left[i\frac{M_W}{v}\cos(\beta-\alpha)\left(W^{-\mu}h\partial_\mu H^+ - W^{-\mu}H^+\partial_\mu h\right) + \text{h.c}\right]\\ &+ \left[\frac{M_W}{v}\left(W^{-\mu}A\partial_\mu H^+ - W^{-\mu}H^+\partial_\mu A\right) + \text{h.c}\right].\end{aligned} \tag{17}$$

We choose the most canonical 2HDM, that is the Type-II 2HDM for our analyses. In a Type-II framework, the Yukawa terms look like

$$\mathcal{L}_h^Y = -\frac{1}{v}\left[\frac{\cos\alpha}{\sin\beta}\sum_{q_u} y_{m_{q_u}} q_u\bar{q}_u h + \frac{\sin\alpha}{\cos\beta}\sum_{q_d} y_{m_{q_d}} q_d\bar{q}_d h\right], \tag{18}$$

$$\mathcal{L}_H^Y = -\frac{1}{v}\left[\frac{\sin\alpha}{\sin\beta}\sum_{q_u} y_{m_{q_u}} q_u\bar{q}_u H + \frac{\cos\alpha}{\cos\beta}\sum_{q_d} y_{m_{q_d}} q_d\bar{q}_d H\right], \tag{19}$$

$$\mathcal{L}_A^Y = -\frac{i}{v}\left[\cot\beta\sum_{q_u} y_{m_{q_u}} q_u\gamma_5\bar{q}_u A + \tan\beta\sum_{q_d} y_{m_{q_d}} q_d\gamma_5\bar{q}_d A\right], \tag{20}$$

$$\begin{aligned}\mathcal{L}_{H^\pm}^Y =& \frac{1}{2}\Big[\left(-y_{ut}\cos\beta + y_{ub}\sin\beta\right)\left(\bar{t}bH^+ + \bar{b}tH^-\right)\\ &+ \left(y_{ut}\cos\beta + y_{ub}\sin\beta\right)\left(\bar{t}\gamma_5 bH^+ - \bar{b}\gamma_5 tH^-\right)\Big],\end{aligned} \tag{21}$$

with $y_{ut} = \sqrt{2}y_{m_t}/(v\sin\beta)$ and $y_{ub} = \sqrt{2}y_{m_b}/(v\cos\beta)$. The relevant trilinear scalar interactions are part of the Lagrangian $\mathcal{L}_{\phi\phi\phi}$,

$$\mathcal{L}_{\phi\phi\phi} = -v\lambda_{hH^+H^-}hH^+H^- - v\lambda_{hH^+H^-}HH^+H^- - \frac{1}{2}v\lambda_{Hhh}Hh^2, \tag{22}$$

where the couplings have the following expressions:

$$\lambda_{hH^+H^-} = \frac{-1}{2v^2\sin(2\beta)}\Big[m_h^2\cos(\alpha-3\beta)+3m_h^2\cos(\alpha+\beta)-4m_{H^\pm}^2\sin(2\beta)\sin(\alpha-\beta)$$
$$-4M^2\cos(\alpha+\beta)\Big], \tag{23}$$

$$\lambda_{HH^+H^-} = \frac{-1}{2v^2\sin(2\beta)}\Big[m_H^2\sin(\alpha-3\beta)+3m_h^2\sin(\alpha+\beta)+4m_{H^\pm}^2\sin(2\beta)\cos(\alpha-\beta)$$
$$-4M^2\sin(\alpha+\beta)\Big], \tag{24}$$

$$\lambda_{Hhh} = \frac{-1}{2v^2\sin(2\beta)}\Big[(2m_h^2+m_H^2)\cos(\alpha-\beta)\sin(2\alpha)$$
$$-M^2\cos(\alpha-\beta)(3\sin(2\alpha)-\sin(2\beta))\Big]. \tag{25}$$

Here M^2 is the shorthand notation for $m_{12}^2/(\sin\beta\cos\beta)$. This completes the whole discussion on the formalism of a specific 2HDM, which could further be used for most of the phenomenological studies to explain the experimental data; and vice-versa from experimental data the parameters of this model can be constrained and hence search studies will follow.

In several places we discussed an additional particle χ which could be characterised as a DM candidate scalar or a scalar or maybe vector-like fermions. Lets consider the case when χ is a DM candidate, then the 2HDM model potential defined in Eq. (2) needs an additional term. One can consider χ as a gauge-singlet scalar, as a DM candidate if its mixing with the doublets Φ_1 and Φ_2 can be prevented by the introduction of some discrete symmetry. One such symmetry is a Z_2 under which χ is odd and all other fields are even. This also ensures the stability of χ. Thus the most general potential consistent with the gauge and Z_2 symmetries can be written as:

$$V(\Phi_1,\Phi_2,\chi) = V(\Phi_1,\Phi_2)+\frac{1}{2}m_\chi^2\chi^2+\frac{\lambda_{\chi 1}}{2}\Phi_1^\dagger\Phi_1\chi^2+\frac{\lambda_{\chi 2}}{2}\Phi_2^\dagger\Phi_2\chi^2+\frac{\lambda_{\chi 3}}{4}(\Phi_1^\dagger\Phi_2+\text{h.c})\chi^2. \tag{26}$$

And hence the complete interaction terms with h and H will follow as

$$\mathcal{L}_{\phi\chi\chi} = -\frac{1}{2}v\lambda_{h\chi\chi}h\chi^2-\frac{1}{2}v\lambda_{H\chi\chi}H\chi^2, \tag{27}$$

where the couplings are given as,

$$\lambda_{h\chi\chi} = \lambda_{\chi_1}\cos\beta\sin\alpha-\lambda_{\chi_2}\sin\beta\cos\alpha-\frac{1}{2}\lambda_{\chi_3}\cos(\beta+\alpha), \tag{28}$$

$$\lambda_{H\chi\chi} = -\lambda_{\chi_1}\cos\beta\cos\alpha-\lambda_{\chi_2}\sin\beta\sin\alpha-\frac{1}{2}\lambda_{\chi_3}\sin(\beta+\alpha). \tag{29}$$

Similarly, we follow the studies for other characteristics of χ based on gauge and symmetry issues. For most of the details on 2HDMs we refer to Refs. [1, 2].

4. Summary

We briefly reviewed Higgs boson physics with an extension of an additional Higgs-boson-doublet in the SM. Also we talked about a study where two hypothetical scalars H and χ with $2m_h < m_H < 2m_t$ and $m_\chi \approx m_h/2$, were introduced to study the p_T spectrum of h, via a process $pp \to H \to h\chi\chi$. The required vertices are defined in an effective way where $H \to h\chi\chi$ branching fraction may be assumed to be large or small depending on the choice of values of the coupling $Hh\chi\chi$. The introduction of a new scalar, S, has been discussed to explain large branching ratios for $H \to h\chi\chi$ decays with an effective hHS vertex through a $S \to \chi\chi$ decay mode, with $m_h \lesssim m_S \lesssim m_H - m_h$ and $m_S > 2m_\chi$. The introduction of a HSS vertex leads to the $H \to SS$ decay.

We discussed salient features of phenomenological aspects of CP-odd scalar, A, in the 2HDMs. Various channels for phenomenological interests in multi-lepton final sates with same and opposite-sign lepton selection with Z+jets+MET, have been discussed. In this respect we discussed how the associated parameters with the charged Higgs $H^{\pm}$ under prominent decay modes to $W^{\pm}H, H \to h\chi\chi$ and $t\bar{b}$ will be probed at the LHC.

It was discussed that χ need not always be taken as a DM candidate, and its characteristic for different phenomenology may need to be characterised as a scalar or as vector-like fermions.

A formalism was presented in a particular type of 2HDM. An appropriate Lagrangian for this model was presented in detail, which further used various studies, as discussed, for the phenomenology of extra scalars. Considering χ to be a DM candidate, interactions terms in a specific 2HDM scenario is presented.

Acknowledgements

T.M. is supported by funding from the Carl Trygger Foundation under contract CTS-14:206 and the Swedish Research Council under contract 621-2011-5107.

References

[1] J. F. Gunion, H. E. Haber, G. Kane and S. Dawson, "The Higgs Hunter's Guide", ABP, PERSEUS PUBLISHING, Cambridge, Massachusetts.

[2] G. C. Branco, P. M. Ferreira, L. Lavoura, M. N. Rebelo, M. Sher and J. P. Silva, Phys. Rept. **516**, 1 (2012) [arXiv:1106.0034 [hep-ph]].

IOP Conf. Series: Journal of Physics: Conf. Series **802** (2017) 012008 doi:10.1088/1742-6596/802/1/012008

Neutron irradiation and damage assessment of plastic scintillators of the Tile Calorimeter

J E Mdhluli[a,1], **B Mellado**[a] **and E Sideras-Haddad**[a]

[a] University of the Witwatersrand, 1 Jan Smuts Avenue, Braamfontein 2000, Johannesburg.

E-mail: [1]joyemmie@gmail.com

Abstract. Following the comparative study of proton induced radiation damage on various plastic scintillator samples from the ATLAS-CERN detector, a study on neutron irradiation and damage assessment on the same type of samples will be conducted. The samples will be irradiated with different dose rates of neutrons produced in favourable nuclear reactions using a radiofrequency linear particle accelerator as well as from the SAFARI nuclear reactor at NECSA. The MCNP 5 code will be utilized in simulating the neutron transport for determining the dose rate. Light transmission and light yield tests will be performed in order to assess the radiation damage on the scintillators. In addition, Raman spectroscopy and Electron Paramagnetic Resonance (EPR) analysis will be used to characterize the samples after irradiation. The project aims to extent these studies to include radiation assessment damage of any component that processes the scintillating light and deteriorates the quantum efficiency of the Tilecal detector, namely, photomultiplier tubes, wavelength shifting optical fibres and the readout electronics. They will also be exposed to neutron irradiation and the damage assessed in the same manner.

1. Introduction

The ATLAS detector is used in particle physics experiments that are involved in the search for new particles through high energy proton-proton collisions at the Large Hadron Collider (LHC) of CERN. The tile calorimeter is part of the ATLAS detector, it is the hadronic calorimeter responsible for detecting hadrons, taus, and jets of quarks and gluons. The tile calorimeter consists of a central barrel and 2 extended barrels. Each barrel contains 64 modules that consists of a matrix of steel plates and plastic scintillators. The steel plates act as an absorber medium that converts the incoming jets into a "shower" of particles. The plastic scintillator tiles then absorb the energy of the particles and fluorescence to emit light. The light from the scintillators is passed through wavelength shifting optical fibres and is detected by photomultiplier tubes. The signal is further processed using readout electronics in order to digitize the data for further analysis [1] [2].

Between the central barrel and extended barrels there is what is referred to as the Gap region. This region contains additional plastic scintillators that are radially distributed within the region. During the first run of data taking, the scintillators in the gap region were exposed to a radiation environment of up to 10kGy/year. It is predicted that during the high luminosity (HL)-LHC run time, the scintillators in the Gap region will sustain a significantly large amount of radiation damage and will require replacement during the 2018 upgrade. This prediction has led to the comparative study of proton induced radiation damage on plastic scintillators conducted by H Jivan [1] [3], C Pelwan [4] and S Liao [5].

To further extend the project on the comparative study, neutron irradiation and damage assessment studies on the same plastic scintillator samples will be conducted. The study will follow the same experimental procedures as that of the proton induced radiation damage study.

2. Scintillation mechanism

The plastic scintillator samples that will be studied are organic scintillators, these scintillators have a basic scintillation mechanism that involves Foster energy transfer and self-absorption. They consists of one or two dopants [6]. The scintillation mechanism of organic scintillators is determined by the physics and chemistry of the benzene ring. An organic scintillator scintillates despite its crystal form, whether it be it a liquid, a gas or imbedded in a polymer. The chemical bonds found within a benzene ring are: σ-bonds that are in the plane with bond angle 120° and are from $sp3$ hybridization. The other chemical bonds found are π-orbitals which are out of plane and overlap. The π-electrons are completely delocalized.

Looking at the scintillation mechanism after the scintillator has absorbed the photon or excitation by ionization, the molecule will undergo vibrational relaxation to the $S10$ state. The $S10$ excited state radiatively decays to the vibrational sub-levels of the ground state. The lifetime of the $S10$ state is in the nanoseconds time range. The short lifetime allows for the fluorescence emission spectrum to be roughly a "mirror image" of the absorption spectrum, in other words, they have the same spacing. The emitted photons have less energy than the $S00 - S10$ phase transition and that's where the important Stokes shift is observed. There is no $S2 - S0$ emission, thus there is an internal non-radiatively de-excitation occurring within the scintillator taking place in the picoseconds time range. The excited triplet state cannot decay to the ground state as a result angular momentum selection rules, it therefore results in a delayed fluorescence and phosphorescence [7].

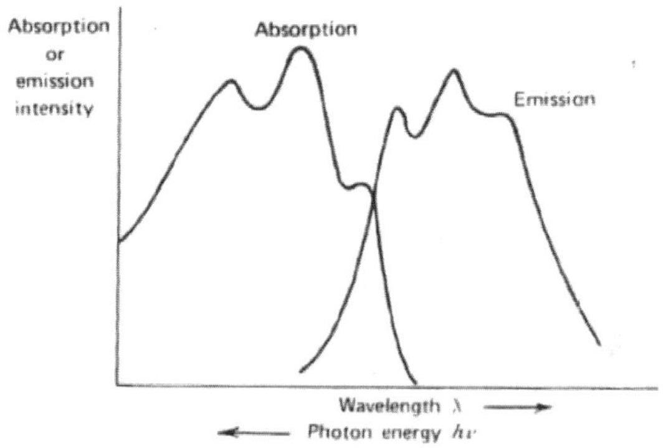

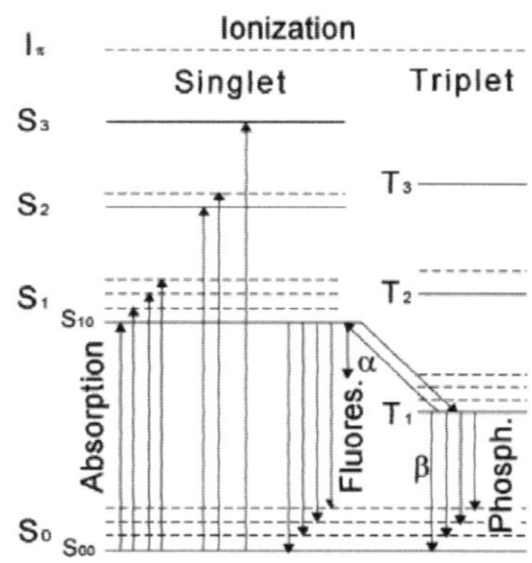

Figure 1. A fluorescence absorption and emission spectra (left) and scintillation mechanism for and organic scintillator (right) [7].

3. Neutron physics

3.1. Neutron production

Neutrons can interact with both nuclei and electrons in matter, with the scattering cross-section of similar sizes for both. Since they can penetrate deeply into matter, they provide an ideal probe of bulk properties of matter which is ideal for our plastic scintillator study [8]. There are 2 fundamental mechanisms that can provide neutrons for slow-neutron scattering purposes in present day research facilities, namely, fission and spallation.

In nuclear physics, nuclear fission can either be a radioactive decay process or a nuclear reaction where a nucleus of an atom will split into smaller parts of lighter nuclei. In an induced nuclear fission reaction, a slow/thermal neutron is absorbed by a nucleus; Uranium-235 for example. The nucleus is turned into an excited Uranium-236 where the excitation energy is provided by the kinetic energy of the neutron plus the forces that bind the neutron. The Uranium-236 then splits into fast moving lighter nuclei/elements and release 2-3 free neutrons. "Prompt gamma rays" are produced concurrently with the free neutrons [9].

The second mechanism used as a neutron source is spallation. Spallation is a process where fragments of a material are ejected from a body due to stress or impact. A spallation source produces pulsed or quasi-continuous neutron beams through the acceleration of protons hitting a target material of a heavy nuclei and releasing neutrons. These heavy nuclei include lead, mercury, tantalum and etc. In contrast to the above mentioned mechanism, one spallation reaction releases about 20-30 neutrons per incident proton particle.

Neutron radiation is often referred to as indirect ionizing radiation. Unlike charged particles that ionize materials through electron excitation, neutrons have no charge and therefore cannot excite electrons. Despite their indirect ionization, neutrons are largely ionizing and this is best observed from the following example: when a neutron is absorbed and results into gamma ray emission, the gamma ray subsequently removes an electron from an atom or a nucleus recoiling from a neutron interaction is ionized and therefore causing more traditional subsequent ionization in other atoms. Since neutrons are uncharged, they are more penetrating than alpha and beta radiation. Neutron radiation from fission can be achieved through an effective fission chain reaction. The neutrons produced during fission should only be captured from a fissionable nuclei which then in turn splits and releases more neutrons [9].

What effects does neutron radiation have on materials? High-energy neutrons will not only degrade and damage materials but neutron bombardment to materials will create collision cascades. These cascades produce point defects and dislocations within the material which are responsible for the microstructural changes occurring over time to materials that are exposed to radiation. The radiation damage in the material occurs as a result of the interaction of an energetic particle with a lattice atom within a material. The collision causes a significantly large amount of kinetic energy transfer to the lattice atom from the neutron, the atom is then displaced from its lattice site and becomes what is known as a primary knock-on atom (PKA). As the PKA's collide with each other, they lose energy with each collision and terminate as interstitials; and effectively creating a series of Frenkel defects within the lattice. Another consequence of the collisions is heat, this is created from the electronic energy loss. The magnitude of the damage caused by a single 1 MeV of a neutron creating a PKA in an ion lattice is such that it will produce approximately 1100 Frenkel pairs. This entire cascade of events happens over a very small time-scale of 10^{-13} seconds and can therefore not be observed without the aid of computer simulations [9].

3.2. The Monte-Carlo N-Particle 5 code

The Monte Carlo method is used to simulate statistical processes theoretically, in particular complex problems that cannot be solved/modelled using computer codes that use deterministic methods. The Monte Carlo N-Particle (MCNP) code is used to simulate neutron, photon and electron or coupled neutron/photon/electron transport. The code is in 3-D and is capable of tracking up to 34 particles and 4 light ions [10]. The code takes into account the absorption and the moderation of the neutrons as they travel though matter.

Considering a beam of neutrons passing through bulk matter, the intensity of the beam will decrease as the beam travels through the material as neutrons are being removed by nuclear reactions. Slow thermal neutrons disappear through capture in the form of (n, γ) reactions. The cross section of capture reactions are often dominated by one more resonances causing the cross section to become very large. When the reactions are off resonance, the cross section decreases with increasing velocity with the relationship: v^{-1}. The neutrons slow down and become moderated due to the elastic and inelastic scattering processes that occur as the cross section decreases therefore there is a higher probability of absorption to occur. The probability for resonant or non-resonant absorption to occur for neutrons is high in the eV energy range, this is the energy range neutrons are in after undergoing scattering processes from the initial energy of 1 MeV.

The loss in intensity of the neutron beam is given by:

$$dI = -I\sigma_t n dx \tag{1}$$

where σ_t is the total cross section that includes the scattering process, ndx is the atoms per unit surface of the beam or material that the neutrons will encounter, n is the number of atoms per unit volume of the material and dx is the thickness of the material. The intensity will decrease with absorber thickness according to an exponential relationship given by:

$$I = I_0 e^{\sigma_t n x} \tag{2}$$

Considering an elastic collision between a neutron with initial energy E and velocity v with a target atom with mass A that is initially at rest, then applying the laws of conservation of energy and linear momentum will give the following ratio between the final neutron energy E' and initial energy:

$$\frac{E'}{E} = \frac{A^2 + 1 + 2A\cos\theta}{(A+1)^2} \tag{3}$$

where θ is the scattering angle in the centre-of-mass system. When there is no scattering occurring, the ratio will be given by $E'/E = 1$. The maximum energy loss occurs during a head on collision, when $\theta = 180^o$ and can be expressed as:

$$\left(\frac{E'}{E}\right)_{min} = \left(\frac{A-1}{A+1}\right)^2 \tag{4}$$

These expressions only apply for a single scattering event but since each neutron will scatter more than once, the energy loss has to be calculated repeatedly. The MCNP code does this calculation, a detailed discussion is found in S Krane [12].

The calculations can be made more quantitative by defining a parameter ξ to represent the average value of $log(E'/E)$ after a single collision, this parameter can be expressed as:

$$\xi = \left[log\frac{E'}{E}\right]_{av} \tag{5}$$

$$= \frac{\int \left[\frac{(A+1)^2}{A^2+1+2A\cos\theta}\right] d\Omega}{\int d\Omega} \tag{6}$$

where $d\Omega$ is the element of solid angle in the centre-of-mass system. Assuming that the scattering is isotropic then the integration can be carried out and give:

$$\xi = 1 - \frac{(A-1)^2}{2A} log\frac{A-1}{A+1} \tag{7}$$

The average value of $logE'$ is decreased after each collision by an amount the parameter represents and therefore after n collisions, the average value of $logE'$ will be $logE'_n$ and given by:

$$logE'_n \simeq logE - n\xi \tag{8}$$

This equation follows equation (5) above.

The calculations above are a good approximation for neutrons in the MeV energy range but as thermal temperatures are approached, the thermal motion of the atoms of the moderator are comparable to the speeds of the neutrons and the above calculations cannot account for their energy loss. A statistical mechanics approach to the problem will give a better approximation. Assuming that after sufficient time the neutrons will reach a thermal equilibrium with the moderator at a certain temperature then the neutrons can be described by a Maxwellian speed distribution given by:

$$f(v)dv = 4\pi n \left(\frac{m}{2\pi kT}\right)^{2/3} v^2 e^{-mv^2/2kT} dv \tag{9}$$

where $f(v)dv$ gives the fraction of neutrons with speeds v and $v + dv$, m is the mass of the neutron and n is the total number of neutrons per unit volume. The above expression can be written in terms of energy, giving [11]:

$$f(E)dE = \frac{2\pi n}{(\pi kT)^{3/2}} E^{1/2} e^{-E/kT} dE \tag{10}$$

4. Experimental procedure

The same experimental procedure followed in the proton irradiation study will be followed in this study too, with a few minor changes. The same plastic scintillator samples, polyvinyl toluene (pvt) and polystyrene (ps) based plastic scintillators purchased from ELJEN technologies and Saint Gobain Crystals will be studied. In addition to commercial scintillators, standard tile calorimeter plastic scintillators manufactured at CERN will also be studied. Instead of using the stopping range of ions in matter (SRIM) program to simulate the damage caused by the neutrons, the MCNP 5 code will be used to simulate the cascade event and calculate the energy loss and dose rate. The samples will be cut and polished in our sample preparation lab located within the Physics department. The samples will be irradiated with different dose rates of neutrons produced in favourable nuclear reactions using a radiofrequency linear particle accelerator as well as from the SAFARI nuclear reactor at NECSA and at the Joint Institute of Nuclear Research (JINR) in Dubna. Light transmission and light yield tests will be performed in order to assess the radiation damage on the scintillators. In addition, Raman spectroscopy and Electron Paramagnetic Resonance (EPR) analysis will be used to characterize the samples after irradiation and compare those results to un-irradiated samples.

5. Additional work

The project aims to extent these studies to include radiation assessment damage of any component that processes the scintillating light and deteriorates the quantum efficiency of the Tilecal detector, namely, photomultiplier tubes (PMT), wavelength shifting optical fibres and the readout electronics. The light that is emitted by the scintillator and passed through wavelength shifting optical fibres is detected by the PMT. The light that enters the PMT is detected and produces an output signal through a series of processes. The light passes through the input window to excite electrons in the photocathode such that photo-electrons are emitted into the vacuum. The photoelectrons are then accelerated and focused by focusing an electrode onto the first dynode where they are multiplied from secondary electron emission. The secondary electron emission is repeated at each of the successive dynodes. The multiplied secondary electrons that are emitted from the last dynode are finally collected by the anode [13].

PMTs will allow for a well conducted comparative study to determine which dynode material is more efficient and sustains the least damage from radiation. Major secondary emissive materials used for most dynodes are alkali antimode, beryllium oxide (BeO), magnesium oxide (MgO), gallium phosphide (GaP) and gallium arsenic phosphide (GaAsP), these will be some

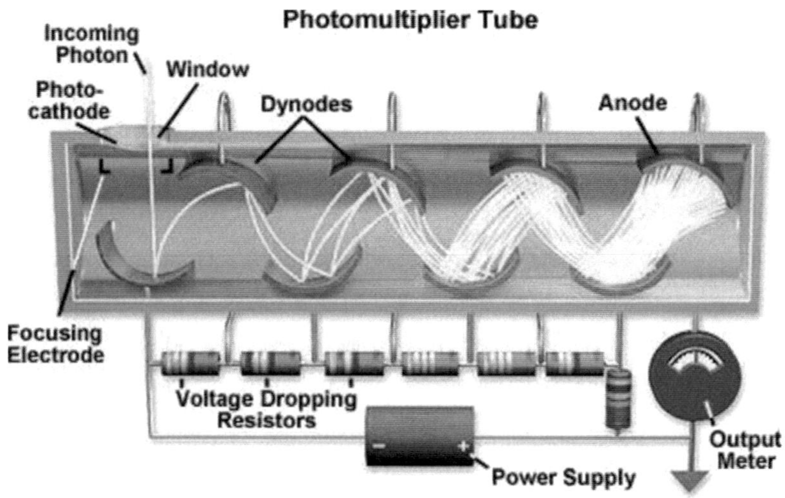

Figure 2. Construction of a photomultiplier tube [12].

of the material intended to be studied in the project [13]. The same neutron irradiation and damage assessment procedures followed for the plastic scintillators will apply to the dynode materials.

References

[1] Jivan H, Erasmus R, Mellado B, Peters G, Sekonya K, Sideras-Haddad E 2014 Radiation hardness of plastic scintillators for the Tile Calorimeter of the ATLAS detector (Submitted to the SAIP conference)
[2] Succurro A 2012 *Physcis Procedia* **37** 229-237
[3] Jivan H, Mellado B, Sideras-Haddad E, Erasmus R, Liao S, Madhuku M, Peters G, Solvyanov O 2015 *Journal of Physics conference series* **645** 012019
[4] Pelwan C, Jivan H, Joubert D, Keartland J, Liao S, Peters G, Sideras-Haddad E 2015 *Journal of Physics conference series* **10** 645:012023
[5] Liao S, Erasmus R, Jivan H, Pelwan C, Peters G, Sideras-Haddad E 2015 *Journal of Physics conference series* **10** 645:012021
[6] D'Ambrosio C, Gys T *A short overview of organic scintillators* [Online] Available: *http://ph-dep-dt2.web.cern.ch/ph-dep-dt2/CAT2005₃a.pdf* [Accessed 15 02 2016]
[7] Chen M 2011 *Queen's University, PHYS 352: Measurement, Instrumentation and Experiment Design* [Online] Available: *www.physics.queensu.ca/ phys352/lect19.pdf* [Accessed 15 11 2016]
[8] Holbert H E *Radiation Effects and Damage* [Online] Available *http://holbert.faculty.asu.edu/eee560/RadiationEffectsDamage.pdf* [Accessed 30 01 2016]
[9] Bisanti P 1986 *Magnetic properties of matter* (Turin:World Scientific) ed Borsa F, Tognetti V p 385-408
[10] Monte Carlo Group: Los Alamos National Laboratory [Online] Available: *https://mcnp.lanl.gov/* [Accessed 24 11 2016]
[11] Krane S K 1987 *Introductory to nuclear physics* (New York, Chichester, Brisbane, Singapore, and Toronto: John Wiley & Sons) p 445- 451
[12] Image downloaded from: *https://micro.magnet.fsu.edu/primer/digitalimaging/concepts/photomultipliers.htm* [Accessed 25 02 2016]
[13] Hamamatsu 2007 *Photomultiplier Tubes: Basics and Applications* [Online] Available: *https://www.hamamatsu.com/resources/pdf/etd/*$PMT_handbool_v3aE - Chater2.pdf$ [Accessed 30 01 2016]

IOP Conf. Series: Journal of Physics: Conf. Series **802** (2017) 012009 doi:10.1088/1742-6596/802/1/012009

Energy density of light quark jet using AdS/CFT

R Morad[1] **and W A Horowitz**[2]

Department of Physics, University of Cape Town, Private Bag X3, Rondebosch 7701, South Africa

E-mail: [1]razieh.morad@uct.ac.za, [2] wa.horowitz@uct.ac.za

Abstract.
We study the energy loss rate of light quarks via the AdS/CFT correspondence in both a static and an expanding plasma. Unlike heavy quarks, light quark energy loss in AdS/CFT is surprisingly dependent on both the string initial conditions and the very definition of the jet itself in the gravity theory. We aim to more closely match the string initial conditions to those expected from perturbative quantum chromodyanics (pQCD)–the theory known to describe the physics of high-momentum particles at early times in heavy ion collisions–by computing the energy-momentum tensor associated with the propagation of the classical string solution. The jet energy-momentum tensor in a strongly-coupled calculation can be found by a superposition of contributions from a collection of point particles whose paths approximate the evolution of the string world-sheet. My results show that some times after creation the pair of quark-anti quark, the energy density is not time dependent. This means that the corresponding jet does not lose energy and the associated nuclear modification factor would be one as expected. Also, the results reveal the virtuality dependency of energy density distribution over space. As expected, the energy of a more virtual jet is spread over wider angles.

1. Introduction

The spectacular measurements from the Relativistic Heavy Ion Collider (RHIC) and the Large Hadron Collider (LHC) provide compelling evidence that the matter produced in heavy ion collision is a deconfined state of QCD, Quark-Gluon Plasma (QGP) [1, 2, 3, 4], at temperatures above ~160 MeV which appears to be nearly perfect, with an extremely low viscosity-to-entropy ratio $\eta/s \sim 1/4\pi$ [5, 6]. Jets are produced within the expanding fireball and probe the QGP. Analyses the energy loss of these energetic partons as they travel throw QGP may reveal extremely valuable information about the dynamics of the plasma and exhibit distinctive properties such as jet-quenching which can clearly be observed at RHIC [1, 2, 3, 4] and more recently LHC [7, 8, 9].

While lattice QCD is the proper tool for understanding the static equilibrium thermodynamics of such strongly coupled plasma, it does not allow us to calculate its dynamics evolution on heavy-ion collision. Recently, a novel tool called "the AdS/CFT correspondence" [10, 11, 12, 13, 14] provide valuable insight into the strongly coupled plasma. In this paper, we review the string setup corresponding to a light quark jet in AdS/CFT and the nuclear modification factor comes from this setup. The result demonstrates the dynamics of jet depends on the initial conditions of the string profile [15]. It is obviouse that any further progress in underestanding the jet in AdS/CFT needs a better underestanding of the string initial conditions. The only way to constraint the string initial conditions is calculation of a gauge invariant quantity in the field

IOP Conf. Series: Journal of Physics: Conf. Series **802** (2017) 012009 doi:10.1088/1742-6596/802/1/012009

theory which is stress tensor. We obtain the energy density of light quark jet in the vacuum using the null string approximation in AdS_5.

2. Light quark jet energy loss

According to the AdS/CFT correspondence [13], the $\mathcal{N} = 4$ SYM theory in the large N_c and large 't Hooft coupling is dual to classical supergravity on ten-dimensional $AdS_5 \times S^5$ geometry [10]. In order to study the theory at finite temperature, one can add black hole (BH) to the geometry [11] which yields to the AdS-Sch metric

$$ds^2 = \frac{L^2}{u^2}\left[-f(u)\,dt^2 + d\mathbf{x}^2 + \frac{du^2}{f(u)}\right], \tag{1}$$

where $f(u) \equiv 1 - (u/u_h)^4$ is the blackening factor and L is the AdS curvature radius. Four dimensional Minkowski coordinates are denoted by x_μ and the coordinate u is an inverse radial coordinate. So, the boundary of the AdS-Sch spacetime is at $u = 0$ and the event horizon is located at $u = u_h$. The temperature of the equilibrium SYM plasma relates to the event horizon as $T \equiv \frac{1}{(\pi u_h)}$. Fundamental quarks are described by open strings attached to the D7 flavor branes. These branes extend along the radial coordinate from the boundary at $u = 0$ down to maximal coordinate at $u = u_m$ as well as fill the whole 4d Minkowski space. Also, they wrap on S^3 from the S^5 sphere. The bare mass M of quark is proportional to $1/u_m$ [16], so for massless quarks the D7 brane should fill the whole radial direction. In the $5d$ geometry these strings can fall unimpeded toward the event horizon until their end points reach the radial coordinate u_m where the D7 brane ends. Since for sufficiently light or massless quarks, open string end points can fall into the horizon. In order to study the back-to-back jets, we consider the configurations in which the two endpoints of string move away from each other as the total spatial momentum of the string vanishes. By choosing the appropriate frame, one half of the string has a large spatial momentum in x direction, and the other half of the string carries a large negative spatial momentum. We will limit our attention to strings which create at a point (x_c, u_c) at initial time $t_c = 0$, and then move in one direction in the R^3 space, x direction. By time evolution, the string evolves from a point into an extended object and the string endpoints fall toward the horizon.

The dynamics of string is governed by the Polyakov action

$$S_P = -\frac{T_0}{2}\int d^2\sigma\,\sqrt{-\eta}\,\eta^{ab}\,\partial_a X^\mu \partial_b X^\nu\, G_{\mu\nu}\,. \tag{2}$$

Variation of the Polyakov action with respect to the embedding functions X^μ leads to the equation of motion which can be solved numerically by considering the appropriate initial conditions for the string profile.

In order to study jets in AdS/CFT we need to define the proper objects in the dual string theory that corresponds to a jet, a slippery object even in field theory; jets are truly only defined by the algorithm used to measure them. Presumably the ideal way to compute jet observables in AdS/CFT is to compute the energy momentum tensor associated with falling string on the dual field theory by solving the gravitational bulk-to-boundary problem and then"run" a jet finding algorithm on the result. However, there are lots of attempts to define jet in the string theory side itself [17, 15]. The authors of [17] are motivated by the localization of the baryon density on the boundary which is of scale of order $\Delta x \sim 1/\pi T$ and defined jet as a part of string which is in the Δx spatial distance from the endpoint. We called this as the "$\Delta x - prescription$" of jet [17]. In [15], we proposed a jet prescription, Δu prescription, motivated by the separation of energy scales in, e.g., thermal field theory. In our prescription the portion of the string above some cutoff $u = u^*$ in the radial direction is considered part of the jet; the portion of the string

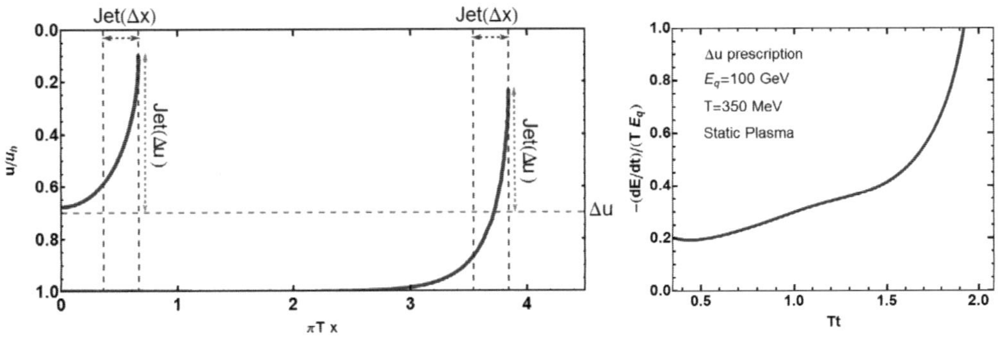

Figure 1. (a) Illustration of the Δx and Δu prescriptions of a jet in the string theory; see text for details. (b) The instantaneous energy loss of a light quark jet as a function of time in the AdS-Sch using the Δu prescription. The normalization constant $E_q = 100$ GeV is the initial energy of the jet, which has a virtuality of 175 GeV^2, and $T = 350$ MeV is the temperature of the plasma.

below the cutoff is considered part of the thermalized medium. By choosing any value of u above the black hole horizon as the cutoff, we regain the natural result that a jet that is thermalized no longer has detectable energy or momentum. We evaluated the energy loss rate of jet using this prescription which shows a Bragg peak at late times figure 1.

Then, we calculated an approximation of the nuclear modification factor R_{AA} for jet using our energy loss model of jet in three different metrics AdS_5, AdS-Sch, and JP metric corresponding to the vacuum, static and expanding plasma respectively. Our results are shown in figure 2. The purple curve shows the R_{AA} obtained from the AdS-Sch metric, while the blue curve is the R_{AA} obtained from the JP metric. We expect that the red line which is the R_{AA} obtained from the AdS_5 would be one since we expect that jets produced in the pp collision do not loss their energy.

Since the definition of R_{AA} measured the difference between the medium and vacuum effects on jet, we define a renormalized R_{AA} in AdS/CFT by dividing the medium R_{AA} by the vacuum R_{AA}, plotted in figure 2(b). Results are in a good agreement with the CMS preliminary data for the most central Pb-Pb collision at $\sqrt{s_{NN}} = 2.76\,GeV$ [18]. In our calculations, we have used the value of t Hooft coupling $\lambda = 5.5$ comes from lattice calculation for quark-anti-quark potential.

3. SYM stress-tensor

According to AdS/CFT dictionary, presence of a source in the bulk perturb the geometry. The behaviour of the metric perturbation near the boundary determine the energy-momentum tensor of jet on the boundary. The metric perturbation can be obtained by solving the linearized Einstein equation. The metric perturbation contains 15 degrees of freedom which couple to each other via the linearized Einstein equations. On the other hand, the 4d SYM stress tensor is traceless and conserved and consequently containes 5 independent degrees of freedom. These two quantities must be contrasted with each other. So, not all of these degrees of freedom in the metric perturbation are physical.

It is possible to construct gauge invariant quantities out of linear combinations of metric perturbations and its derivatives [19] and calculate the gauge invariant variables directly from

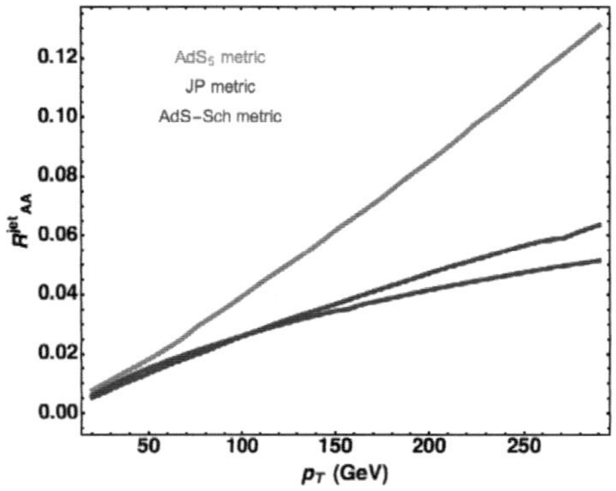

(a)

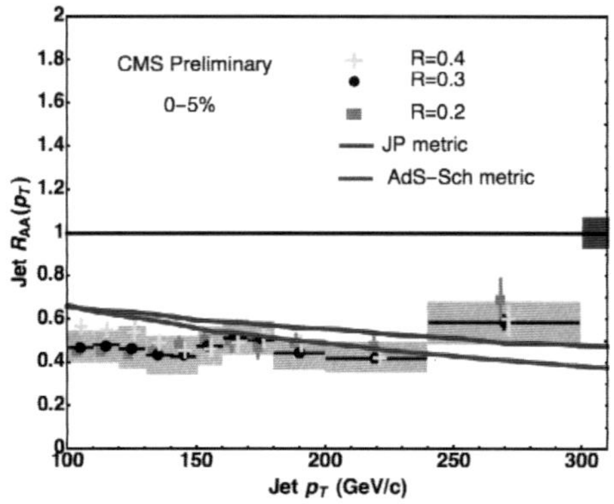

(b)

Figure 2. (a) Jet R_{AA} as a function of p_T in the most central Pb-Pb collision obtained via AdS/CFT in AdS_5 (red), JP (blue) and AdS-Sch (purple) metrics. (b) AdS/CFT jet R_{AA} as a function of p_T compared with the preliminary CMS data in different effective cone sizes for anti-k_T jets using the Bayesian unfolding method for most central Pb-Pb collision at the LHC with $\sqrt{s} = 2.76$ TeV per nucleon [18]. The value of t Hooft coupling used in these calculations is $\lambda = 5.5$.

them. In [20], the energy density of stress tensor on the boundary at a space-time point $(t_b, \mathbf{r_b})$ has been evaluated as

$$\begin{aligned}\mathcal{E}(t_b, \mathbf{r_b}) &= \frac{2L^3}{\pi}\int d^4r\,\frac{du}{u^2}\,\Theta(t_b - t)\,\delta''(W)\,[u(2t_{00} - t_{55}) - (t_b - t)t_{05} + (x_b - x)^i t_{i5}] \\ &+ \frac{2L^3}{3\pi}\int d^4r\,\frac{du}{u}\,\Theta(t_b - t)\,\delta'''(W)\,[|\mathbf{r_b} - \mathbf{r}|^2\,(2t_{00} - 2t_{55} + t_{ii}) - 3(x_b - x)^i\,(x_b - x)^j\,t_{ij}],\end{aligned} \quad (3)$$

where $W = -(t_b - t)^2 + (\mathbf{r_b} - \mathbf{r})^2 + u^2$ is the bulk to bounadry propagator and t_{MN} is the bulk stress tensor. At time t, the bulk excitation localized at $(t, \mathbf{r})$ emits a gravitational wave h_{MN} which propagates through AdS_5 at the respective speed of light up to the measurement point $(t_b, \mathbf{r_b})$ on the boundary. The δ function in the integrand represents the support of the retarded bulk-to-boundary propagator for the Einstein equations in AdS_5.

The falling string can be approximated with a null string such that each point of string follows a null geodesic in the bulk with the same velocity as string velocity, after the geometric optic expansion [21]. Then, the jet energy-momentum tensor in a strongly-coupled calculation can be relatively easily found by a superposition of contributions from a collection of point particles whose paths approximate the evolution of the string worldsheet. The point particle energy density in vacuum is obtained [22]

$$\mathcal{E}_{particle}(t_b, \mathbf{r_b}) = \frac{E_0}{2\pi\gamma^4}\frac{(t_b + u_c\gamma)^2}{(t_b + u_c\gamma - v(x_b + u_c v\gamma))^3}\,\delta(\,(t_b + u_c\gamma)^2 - x_{\perp b}^2 - (x_b + u_c v\gamma)^2\,)\,, \quad (4)$$

where x_b is the direction along the motion of string in 4d Minkowski space and $x_{\perp b}$ is the transverse direction. In figure 3, the angular distribution of energy is plotted for two strings with the same energies but different virtualities. As we expected, after a short time after creation the enregy is highly concentrated at $\theta = 0, \pi$ correcponding to the position of the string endpoints and also it is no time dependent. This means that the corresponding jet does not lose energy and the associated nuclear modification factor would be one as expcted. Also, the

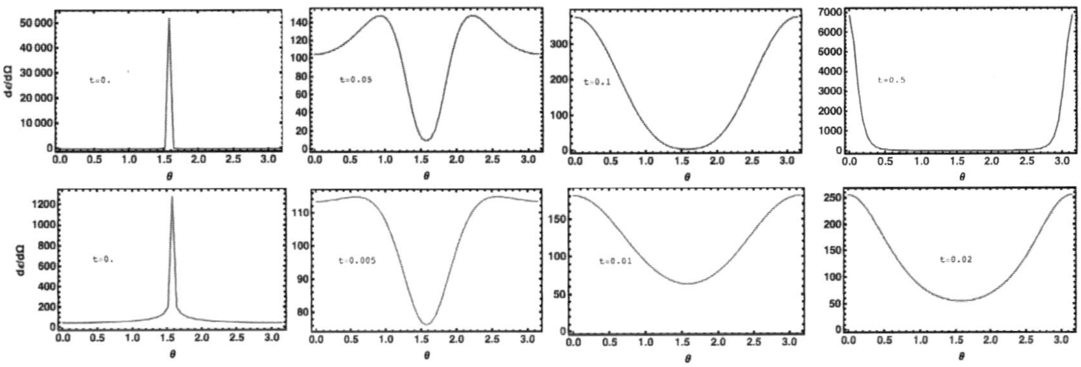

Figure 3. Angular distribution of light quark jet energy for two different jets with 100 GeV initial energy. The first falling string has $Q^2 \simeq 176\,GeV^2$ while the second string has $Q^2 \simeq 6000\,GeV^2$. θ is the angle with the direction of motion of the string endpoint x.

results reveal the virtuality dependency of energy density distribution over space. It should be mentioned that we have used the particle physics sign convention for the virtuality of quark defined as $Q^2 = E_q^2 - p_q^2$. Actually, energy of a more virtual jet spread over a wider angle.

4. Conclusions

We have purposed a novel prescription of jet based on energy scale separation in AdS/CFT. We have shown that using our prescription of jet, the light quark energy loss exhibit the Bragg peak again at late times in both static and expanding plasmas. This late time behavior of jet energy lost implies that after traveling substantial distances through the plasma, the thermalization of light quark ends with a large amount of energy transferring to the plasma. We considered a brick of plasma and calculated the nuclear modification factor of jet in both AdS-Sch and JP metric. We assumed that the temperature of plasma around 350 MeV at AdS-Sch metric and at initial time in JP metric. Our results show an over suppression of jet of order of ten respect to the data. We investigated that it is because of the falling string setup at AdS space. In fact, R_{AA} of jet using the falling string in AdS_5 which is dual to jets in vacuum is not one, even though it is less than one. We introduced a renormalized R_{AA}^{renorm} by dividing the R_{AA} in the medium to the R_{AA} in the vacuum. Surprisingly, our ratio shows good agreement with the experimental data on the R_{AA}^{jet} of most central Pb-Pb collision at LHC figure 2 (b).

Our results on jets using AdS/CFT correspondence reveals two important issues: 1) The dynamics of jet highly depends on the initial conditions of falling string figure 4, while there is no known map between the string initial conditions and parameters of the dual state in the field theory. Further progress in describing experimental results will require significant advances in the understanding of string initial conditions. 2) That the results of our simple model are in such good agreement with data suggests that we attempt to better define the jet in AdS/CFT and constrain the possible string initial conditions. We can likely accomplish both goals by computing the energy-momentum tensor associated with the propagation of the classical string solution. With the energy- momentum tensor in hand, we should be able to compute directly from the string theory the actual quantities measured experimentally.

One can solve Einstein's equations for the perturbation in the $5d$ geometry due to the presence of the string and according to the bulk to boundary map interpret the near boundary behavior of the metric perturbation as the perturbation in the SYM energy-momentum tensor by the

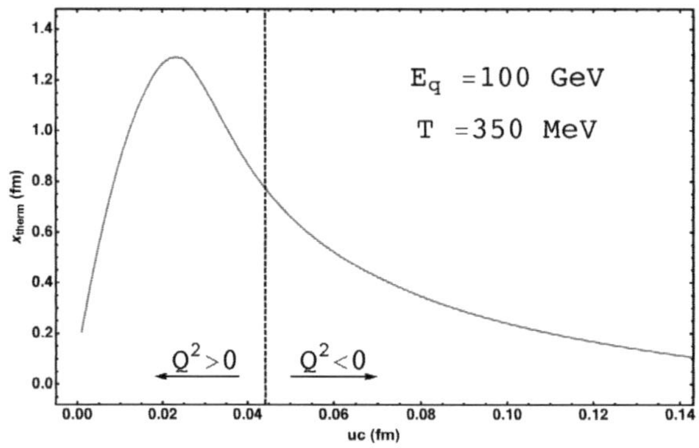

Figure 4. The maximum stopping distance of jet versus the virtuality of quark Q^2. Our result show a nontrivial dependency of the thermalization time to the initial condition of the string.

presence of jet. Our studies show that the energy of jet in vacuum is distributed about the string endpoints on the boundary, see figure 3. Also, after a short time after creation its angular distribution is not time dependent which means that the corresponding jet does not lose energy in vacuum and the associated nuclear modification factor would be one as expcted.

In order to compute the R_{AA} for jet, we need to calculate the boundary energy density of falling string in $AdS - Sch$ and then compare with the energy density in AdS_5 which we left for future work.

References

[1] Adams J *et al.* (STAR) 2005 *Nucl. Phys.* **A757** 102–183 (*Preprint* nucl-ex/0501009)
[2] Adcox K *et al.* (PHENIX) 2005 *Nucl. Phys.* **A757** 184–283 (*Preprint* nucl-ex/0410003)
[3] Arsene I *et al.* (BRAHMS) 2005 *Nucl. Phys.* **A757** 1–27 (*Preprint* nucl-ex/0410020)
[4] Back B B *et al.* 2005 *Nucl. Phys.* **A757** 28–101 (*Preprint* nucl-ex/0410022)
[5] Policastro G, Son D T and Starinets A O 2001 *Phys. Rev. Lett.* **87** 081601 (*Preprint* hep-th/0104066)
[6] Kovtun P, Son D T and Starinets A O 2005 *Phys. Rev. Lett.* **94** 111601 (*Preprint* hep-th/0405231)
[7] Yin Z B (ALICE) 2013 *Acta Phys. Polon. Supp.* **6** 479–484
[8] Aad G *et al.* (ATLAS) 2010 *Phys. Rev. Lett.* **105** 252303 (*Preprint* 1011.6182)
[9] Chatrchyan S *et al.* (CMS) 2011 *Phys. Rev.* **C84** 024906 (*Preprint* 1102.1957)
[10] Maldacena J M 1999 *Int. J. Theor. Phys.* **38** 1113–1133 [Adv. Theor. Math. Phys.2,231(1998)] (*Preprint* hep-th/9711200)
[11] Witten E 1998 *Adv. Theor. Math. Phys.* **2** 253–291 (*Preprint* hep-th/9802150)
[12] Gubser S S, Klebanov I R and Polyakov A M 1998 *Phys. Lett.* **B428** 105–114 (*Preprint* hep-th/9802109)
[13] Aharony O, Gubser S S, Maldacena J M, Ooguri H and Oz Y 2000 *Phys. Rept.* **323** 183–386 (*Preprint* hep-th/9905111)
[14] Casalderrey-Solana J, Liu H, Mateos D, Rajagopal K and Wiedemann U A 2011 (*Preprint* 1101.0618)
[15] Morad R and Horowitz W A 2014 *JHEP* **11** 017 (*Preprint* 1409.7545)
[16] Herzog C P, Karch A, Kovtun P, Kozcaz C and Yaffe L G 2006 *JHEP* **07** 013 (*Preprint* hep-th/0605158)
[17] Chesler P M, Jensen K, Karch A and Yaffe L G 2009 *Phys. Rev.* **D79** 125015 (*Preprint* 0810.1985)
[18] Collaboration C (CMS) 2012 *CMS-PAS-HIN-12-004*.
[19] Kovtun P K and Starinets A O 2005 *Phys. Rev.* **D72** 086009 (*Preprint* hep-th/0506184)
[20] Athanasiou C, Chesler P M, Liu H, Nickel D and Rajagopal K 2010 *Phys. Rev.* **D81** 126001 [Erratum: Phys. Rev.D84,069901(2011)] (*Preprint* 1001.3880)
[21] Chesler P M and Rajagopal K 2014 *Phys. Rev.* **D90** 025033 (*Preprint* 1402.6756)
[22] Hatta Y, Iancu E, Mueller A H and Triantafyllopoulos D N 2011 *JHEP* **02** 065 (*Preprint* 1011.3763)

Non-abelian correction to the Poisson approximation for multi-gluon bremsstrahlung

Andriniaina N. Rasoanaivo and W. A. Horowitz

Department of Physics, University of Cape Town, Private Bag X3, Rondebosch 7701, South Africa.

E-mail: andriniaina@aims.ac.za

Abstract. We present a technique that we are developing to compute the non-abelian effects to the well-known poisson distribution from QED results to obtain the QCD momentum distribution. We first introduce the MHV method used compute scattering amplitudes in gauge theory, especially $\mathcal{N} = 4$ SYM, then we discuss the case of massless QCD. In second part we present why the irreducible representation of the symmetric group is the natural framework to obtain the non-abelian correction to the QED distribution.

1. Introduction

During the last decade, in theoretical and mathematical physics side, one of the most exciting advanced fields of research is the Maximal Helicity Violating technique (MHV) also known as the on-shell method of Quantum Field Theory (QFT). The MHV technique provides a new framework to understand symmetries and its on-shell property makes computation of scattering amplitudes simpler than the standard diagrammatic methods of QFT [1, 2, 3]. In this framework, amplitudes involving a large productions of gluon have been computed for both pure gauge theory and massless QCD.

In the experimental side, the discovery of the quark gluon plasma (QGP) at RHIC and LHC from heavy ion collisions provide a new area to understand the phase diagram of QCD [4]. The QGP is a new phase of matter, where the quarks and gluons are free to move relatively independently, and behaves nearly as a perfect fluid [5]. At low-momentum regime the physics of the QGP are best described by strong-coupling physics of the anti-de-Sitter conformal field theory (AdS/CFT) conjecture, however in the high momentum regime they are described by weak-coupling physics of perturbative QCD. In addition to the observed fluid behaviour QGP there are also evidence of jets productions from heavy ion collisions. It is clear that the jets produced evolve in time with medium and due the in-medium interaction, energy loss are expected. It is also known that jets propagating in colored matter lose energy predominantly through medium-induced emission of gluon radiation [6, 7, 8], a jet scatters off color charges in the QGP matter and radiates gluon bremsstrahlung.

In this essay, we want to introduce our idea that using the MHV technique we are able to use QED radiative corrections as a central point in order to obtain radiative gluon bremsstrahlung in a large number of emission. In another word we will add non-abelian correction to the QED result. It has been shown in the QED that the radiative corrections is trivially resummed and exponentiate into a poisson distribution, for the case of QCD many attempts using different approximations as been done in order to be able to exponentiate the radiative corrections. Here, we want to see a smooth resummation as in QED by using the irreducible representation of symmetric group. Before we get into the detail let remind ourselves that for a given amplitude $\mathcal{M}(k_1, \ldots, k_n)$ that have m bremsstrahlung gluons in the soft-collinear limit factorize as

$$\mathcal{M}(k_1, \ldots, k_n) \to \mathcal{M}(k_{m+1}, \ldots, k_n) J(k_1, \ldots, k_m) \tag{1}$$

where J is the radiative corrections also known as soft collinear factor. That function J captured all information on the behaviour of the amplitude near the soft and collinear limit, and our goal is to take the corresponding correction J_{QED} and find the non-abelian correction using MHV.

$$J_{QCD} = J_{QED} + \text{corrections}. \tag{2}$$

2. Short introduction to MHV amplitudes

Scattering amplitudes are the central objects in perturbative quantum field theory to bridge between theoretical to experimental predictions. To compute scattering amplitudes it is very natural to us to use Feynamn diagram which seems to be the right approach since Feynman diagrams make unitarity and locality manifest. In order to make locality manifest we introduce virtual particles, off-shell particles, but some how after we summed all relevant diagrams the final amplitude doesn't depend on those virtual particles [9].

Complications appear when we want to compute amplitudes with large number of particles, because the number of Feynman diagrams tends to grow fast with the number of particles, for example in $g + g \to g + g + g$ we have 25 diagrams at tree level. It turns out that after simplifications and using spinor helicity representation the amplitude can be written in a very simple expression.

Starting by using the spinor representation instead of the usual four vectors representation, a null momentum vector becomes

$$p_\mu \to p_{a\dot{a}} = (\sigma^\mu)_{a\dot{a}} p_\mu = \lambda_a \widetilde{\lambda}_{\dot{a}}, \tag{3}$$

where λ_a and $\widetilde{\lambda}_{\dot{a}}$ are respectively a left and right handed spinors. And since amplitudes are made up with Lorentz invariant objects, we need to introduce two Lorentz invariant products made of those new variables

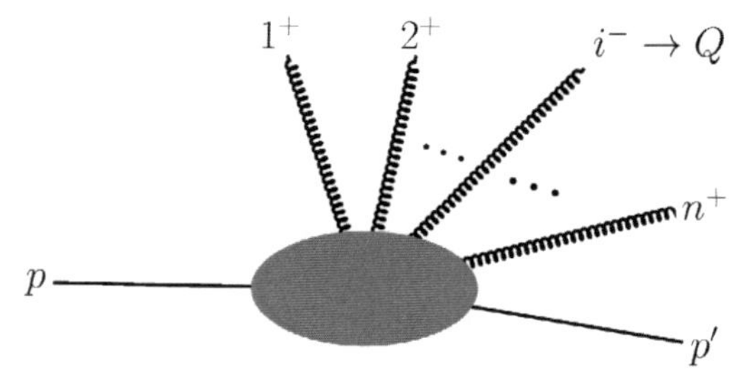

Figure 1. Amplitude corresponding to (6).

$$\langle p, q \rangle = \epsilon_{ab} \lambda_p^a \lambda_q^b \quad \text{and} \quad [p, q] = \epsilon_{\dot{a}\dot{b}} \widetilde{\lambda}_p^{\dot{a}} \widetilde{\lambda}_q^{\dot{b}}. \tag{4}$$

Here ϵ_{ab} and $\epsilon_{\dot{a}\dot{b}}$ are the antisymmetric tensors in two dimensions. The use of left or right handed spinor is now depending on the helicity h of the particles, projection of the spin along the momentum of the particle. In the same way for gauge field ($h = \pm 1$) we can introduce the spinor helicity representation of the polarization vector

$$\varepsilon^-_{a\dot{a}}(p; q) = \frac{\lambda_a \widetilde{\mu}_{\dot{a}}}{[p, q]} \quad \text{and} \quad \varepsilon^+_{a\dot{a}}(p; q) = \frac{\mu_a \widetilde{\lambda}_{\dot{a}}}{\langle p, q \rangle}, \tag{5}$$

with $p = \lambda\widetilde{\lambda}$ and $q = \mu\widetilde{\mu}$. The momentum q is not physical in the sense that it is part of the gauge freedom in the spinor helicity formalism, so $q \neq p$ is an arbitrary momentum. Putting all of this together and using the color-kinematic factorization, the full amplitude for the MHV case, where we have one quark line interact with $n+1$ gluons with positive helicities and one gluon with negative helicity, is given by

$$\mathcal{M}_n = \sum_{\sigma \in S_n} T_{a_{\sigma_1}} \cdots T_{a_{\sigma_n}} A_n(p, p'; \sigma_1, \dots, \sigma_n), \tag{6}$$

where we are summing over all permutation of gluon indices. Here the product of T_{a_i} which are the generators of $SU(N)$ are called the color factor and A_n is called partial amplitude that contain only the kinematics of the process

$$A_n(p, p'; \sigma_1, \cdots, \sigma_n) = \frac{\langle p, Q\rangle^3 \langle p', Q\rangle}{\langle p'p\rangle \prod_{a=1}^{n-1} \langle p, \sigma_a\rangle \langle \sigma_a, \sigma_{a+1}\rangle \langle \sigma_{a+1}, p'\rangle}, \tag{7}$$

were Q is the momentum of the gluon that carry a negative helicity. Now if we consider the case were all positive helicities gluons are bremsstrahlung, and by using the factorisation in near the soft-collinear limit (1) we obtain

$$J_{QCD} = \sum_{\sigma \in S_n} T_{a_{\sigma_1}} \cdots T_{a_{\sigma_n}} \frac{\langle p, Q\rangle \langle Q, p'\rangle}{\prod_{a=1}^{n-1} \langle p, \sigma_a\rangle \langle \sigma_a, \sigma_{a+1}\rangle \langle \sigma_{a+1}, p'\rangle}. \tag{8}$$

3. Non-abelian correction

In this section, we are going to show that the projection of (8) into a different representation will give us the right expansion around the abelian theory. First let us consider the fact that making the gauge field of QCD commute will transform it the theory into QED which consist to transform $SU(N)$ to $U(1)$ or $T_{a_i} \to 1$. Applying such transformation into (8) gives us J_{QED} as expected [10], which is the symmetrization of the kinematic part of J_{QCD}

$$J_{QED} = \sum_{\sigma \in S_n} S_{\sigma(1)\cdots\sigma(n)} = \prod_{I \neq Q} \frac{\langle p, p'\rangle}{\langle p, I\rangle \langle I, p'\rangle}, \tag{9}$$

with

$$S_{\sigma(1)\cdots\sigma(n)} = \frac{\langle p, Q\rangle \langle Q, p'\rangle}{\prod_{a=1}^{n-1} \langle p, \sigma_a\rangle \langle \sigma_a, \sigma_{a+1}\rangle \langle \sigma_{a+1}, p'\rangle}. \tag{10}$$

That lead us to think that the irreducible representation of the permutation group S_n is the mathematical framework that we want to expand the QCD radiative correction since its fully symmetric projection is the QED correspondent.

The irreducible representation of S_n are usually represented by the Young diagrams corresponding. For example for S_2 we have the 2 irreducible decomposition one the symmetric and one antisymmetric, then we can always decompose any tensor of order 2 into symmetric and antisymmetric tensor ($A_{ij} = A_{ij}^{\text{sym}} + A_{ij}^{\text{ant}}$). In a general tensor of order n, we can do a similar decomposition into the irreducible representation of S_n. As presented in [11], we are going to use projectors P_α into the Young diagram correspondent such that

$$\sum_a P_\alpha = 1 \quad \text{and} \quad P_\alpha P_\beta = \delta_{\alpha\beta} P_\alpha. \tag{11}$$

Here we give an example of projectors for $n = 3$:

$$\begin{cases} P_1 = \frac{1}{6}\left[1 + (12) + (13) + (23) + (123) + (132)\right] \\ P_2 = \frac{1}{6}\left[1 - (12) - (13) - (23) + (123) + (132)\right] \\ P_3 = \frac{1}{3}\left[2 - (123) - (132)\right] \end{cases} \tag{12}$$

Using those projection into (8), both for the color factor and for the kinematic part, where it act on the gluon labels $\{1,\ldots,n\}$. Taking $T_{a_1}\cdots T_{a_n} = C_{a_1\cdots a_n}$ for the color factor, and considering the fact that J_{QED} is the symmetric part of the kinematics (9), we obtain

$$J_{QCD} = C^{\text{sym}}_{a_1\cdots a_n} J_{QED}(1,\ldots,n) + \sum_{\alpha\neq 1}\sum_{\sigma\in S_n} C^{(\alpha)}_{a_{\sigma(1)}\cdots a_{\sigma(n)}} S^{(\alpha)}_{\sigma(1)\cdots\sigma(n)}, \tag{13}$$

with $C^{(\alpha)}_{\ldots} = P_\alpha C_{\ldots}$ and similarly for $S^{(\alpha)}_{\ldots}$ are the projected tensor into the symmetry α. This equation looks pretty much close to we expected (2), in the exception we have a color factor with the J_{QED} which make sense since QCD is a theory that carry colors.

4. Conclusion

During the last decade of development of the maximal helicity violating, many progress was made in the understanding of the mathematical structure of the scattering amplitude. In the mathematical physics community those developments push them to go further into deep mathematical framework (twistor theory, positive Grassmannian, Amplituhedron, ...) to understand the very fundamental aspect of QFT. However for the experimental, since the discovery of the quark gluon plasma leads the experimentalists to run many different experiments (p+p, p+Pb, Pb+Pb, ...) in order to test the QCD phase diagram at different scales and conditions.

In this talk, we introduced one of the applications of the actual massless theory into the most exiting research experiments heavy ion collision. This is also an opening into the permutation structure of multi-gluon emission. We saw that the purely symmetric structure correspond to QED, and any antisymmetric structure will be in the non-abelian correction as shown in (13).

In QCD, any excitation from the medium leads to a radiation even if it is a color flip, i.e. no energy loss. However in QED there is no such process as color flip; that means J_{QED} must vanish for $p = p'$, true from (9). So in the limit $p = p'$ the QCD radiative correction is purely non-abelian and our actual work is focussed on understanding the color flip structure which may lead us to the smooth resumation of bremsstrahlung radiave corrections.

Acknowledgments

This work would have been impossible without the support of the Science Faculty PhD Fellowship of UCT. We are also grateful to the supports from SA-CERN and NRF-South Africa.

References

[1] F. Cachazo, P. Svrcek, E. Witten, *"MHV Vertices And Tree Amplitudes In Gauge Theory"*, Journal of High Energy Physics, Volume 2004, JHEP09(2004).
[2] Z. Bern, L. J. Dixon, D. A. Kosower, *"On-shell recurrence relations for one-loop QCD amplitudes"*, Phys. Rev. D 71, 105013, (2005).
[3] A. Brandhuber, B. Spence, G. Travaglini, *"Tree-level formalism"*, Journal of Physics A: Mathematical and Theoretical, Volume 44, Number 45, (2011).
[4] D.H. Rischke, Prog. Part. Nucl. Phys. 52 (2004) 197.
[5] G. Policastro et al., Phys. Rev. Lett. 87, 081601 (2001); P. K. Kovtun et al., ibid.94, 111601 (2005).
[6] Megan Connors, for the ALICE Collaboration, *"Jet Production in p-Pb Collisions"*, arXiv:1409.4655 [nucl-ex].
[7] A. Adare et al., *"Centrality-dependent modification of jet-production rates in deuteron-gold collisions at $\sqrt{s_{NN}}$=200 GeV"*, arXiv:1509.04657 [nucl-ex] (2015).
[8] W.A. Horowitz, *"Heavy Quark Production and Energy Loss"*, Nuclear Physics A, Volumes 904905, Pages 186c193c (2013).
[9] M. E. Peskin, *"Simplifying Multi-Jet QCD Computation"* , arXiv:1101.2414 [hep-ph] (2011).
[10] K. J. Ozeren, W. J. Stirling, *"MHV techniques for QED processes"*, Journal of High Energy Physics, Volume 2005, JHEP11(2005)
[11] T. L. Wade, *"Tensor Algebra and Young's Symmetry Operators"*, American Journal of Mathematics Vol.63, No.3 , pp.645-657, (1941).

Search for resonances decaying to photon pairs in 3.2 fb^{-1} of pp collisions at $\sqrt{s} = 13$ TeV with the ATLAS detector.

Robert Graham Reed
School of Physics, University of the Witwatersrand , Johannesburg 2050, South Africa
E-mail: robert.reed@cern.ch

Abstract. This proceedings summarises the search for resonances decaying into two photons with a mass greater than 200 GeV. Models, which have scalars, such as the extended Higgs sector are used to optimise the analysis. The results presented are for a dataset containing 3.2 fb^{-1} of pp collisions at $\sqrt{s} = 13$ TeV using the ATLAS detector. The data is consistent with the expected background with the most significant deviation appearing in the 750 GeV region with a global significance of 2σ. A detailed description of the analysis will be presented.

1. Introduction
The ATLAS and CMS collaborations first announced the observation of a new scalar boson [1,2] in 2012 marking the discovery of the Higgs boson, the final piece to the Standard Model. The next logical step is the search for physics beyond the Standard Model (BSM). The diphoton channel offers a clean signal, offering excellent mass resolution, and has a well understood background. The LHC is, for the first time, providing means to probe higher energies than ever before. In models with an extended Higgs sector [3–5] new scalars with masses different to the Standard Model (SM) Higgs may be observed. This proceedings describes the analysis that has been optimised for masses larger than 200 GeV [6]. The study involves selecting two photons using tight identification to minimise the background that are not from SM diphoton production. Any Higgs-like resonance will appear as a localised excess above the smooth background.

2. The ATLAS detector
The ATLAS detector [7] is one of two main general purpose detector at the LHC. The detector is cylindrical in shape and has a forward-backward symmetry with a near 4π coverage in solid angle. ATLAS uses a right-handed coordinate system with the origin at the nominal interaction point and the z-axis along the beam pipe with the x-axis pointing towards the centre of the LHC ring. The detector consists of multiple concentric sub detectors. Starting from the inner detector (ID) we have the inner tracker and silicon microstrip detector located inside the transition radiation tracker which provide tracking in the pseudorapidity range $|\eta| < 2.5$, electromagnetic (EM) calorimeter and lead liquid-argon sampling device is divided into one barrel ($|\eta| < 1.45$)) and two end cap regions ($1.375 < |\eta| < 3.2$), the hadron calorimeter is divided into a barrel ($|\eta| < 0.8$) and extended barrel ($0.8 < |\eta| < 1.7$) both of which consist of steel and plastic scintillators.

Each sub detector is designed to identify different particles, their energies and trajectories which when combined together can provide a detailed account of an entire event collision.

3. Data and triggering

The data was recorded by ATLAS during the 2015 data taking period with a centre-of-mass energy of $\sqrt{s} = 13$ TeV with bunch spacing of 25 ns. The average interactions per bunch crossing was about 13 with a peak instantaneous luminosity of $\mathcal{L} = 5.2 \times 10^{33}\ cm^{-2}s^{-1}$. A diphoton trigger with transverse energy (E_T) thresholds of 35 GeV and 25 GeV, for E_T ordered photons, was used. Clusters of energy in the EM calorimeter are reconstructed and required to satisfy a loose criteria to what is expected for EM showers initiated by photons. The efficiency for the signal trigger is close to 99% for events that pass the full selection. Events are only considered if they were recorded during stable beam conditions where the trigger system, tracking devices and calorimeters are all operational. The total integrated luminosity was recorded to be 3.2 fb^{-1}.

4. Monte Carlo simulation

Simulated events are used to profile the diphoton signal shape and to estimate the background by parametrising its shape. The simulation can provide selection inefficiencies as well as detector resolution effects which can be used to correct the results from the data. The signal is simulated as a gluon fusion (ggF) process resulting in a SM Higgs-like boson produced under the same conditions as those expected for data. A single dataset is simulated for various mass hypothesis in the range [200, 2000] GeV. A narrow width approximation (NWA) of 4 MeV is used for the simulated boson which correspond to a 125 GeV SM Higgs boson. The interference between the signal and background is assumed to be tiny and is neglected. Resonances with a large natural width also need to be considered so a large width approximation (LWA) is also studied. To obtain the mass distribution for the large natural width the theoretical line shape is convoluted with a resolution function. The resolution function takes into account the detector and reconstruction effects and takes the form of a double-sided Crystal Ball function (DSCB). The width ($\alpha = \Gamma/m_X$) of the sample is expressed as a percentage of the generated mass hypothesis (m_X). The DSCB is defined as:

$$N \cdot \begin{cases} e^{-t^2/2} & \text{if } -\alpha_{\text{low}} \geq t \geq \alpha_{\text{high}} \\ \dfrac{e^{-\alpha_{\text{low}}^2/2}}{\left[\frac{\alpha_{\text{low}}}{n_{\text{low}}}\left(\frac{n_{\text{low}}}{\alpha_{\text{low}}} - \alpha_{\text{low}} - t\right)\right]^{n_{\text{low}}}} & \text{if } t < -\alpha_{\text{low}} \\ \dfrac{e^{-\alpha_{\text{high}}^2/2}}{\left[\frac{\alpha_{\text{high}}}{n_{\text{high}}}\left(\frac{n_{\text{high}}}{\alpha_{\text{high}}} - \alpha_{\text{high}} + t\right)\right]^{n_{\text{high}}}} & \text{if } t > \alpha_{\text{high}}, \end{cases} \tag{1}$$

where $t = \Delta m_X/\sigma_{\text{CB}}$, $\Delta m_X = m_X - \mu_{\text{CB}}$, N is a normalization parameter, μ_{CB} is the peak of the Gaussian distribution, σ_{CB} represents the width of the Gaussian part of the function, α_{low} (α_{high}) parameterizes the mass value where the distribution switches from a Gaussian to a power-law function on the low (high) mass side, and n_{low} (n_{high}) is the exponent of this power-law function. Large-width samples are produced in the same mass range [200, 2000] GeV as the NWA but with $\alpha = \Gamma/m_X$ ranging from 1% to 10%, in a width range where the interference with the continuous background is expected to be small [8]. Samples obtained with the above procedure for $m_X = 400$ GeV, 800 GeV and 1500 GeV, and $\alpha = 5\%$, 7% and 10%.

5. Object and Event selections

Photon reconstruction begins with EM cells that have a transverse energy exceeding 2.5 GeV. The cells are matched to tracks in the ID and extrapolated to the calorimeter. If no tracks are found then the cluster is marked as a unconverted photon, one track is an electron candidate and two tracks is a converted photon under the hypothesis of $\gamma \rightarrow e^+e^-$. The efficiency to correctly reconstruct photons from the clusters and tracks is 97% with the other 3% incorrectly reconstructed only as electron candidates. Dedicated energy calibrations, for both converted and unconverted photons, are used which account for lateral and longitudinal leakage and other upstream energy losses [9]. Photons are required to satisfy an identification criteria which is based on the shower shape in the EM calorimeter and the energy leakage in to the hadronic calorimeter. The efficiency of the tight identification for unconverted (converted) photons asymptotically reaches 95% (98%) for $E_T > 200$ GeV. In order to reduce the backgrounds from jets an isolation requirement is used on photon candidates. This makes use of the isolation transverse energy (E_T^{iso}) which is the sum of transverse energies of the positive-energy topological clusters within a cone of size $\Delta R = 0.4$, where $\Delta R \equiv \sqrt{(\Delta\eta)^2 + (\Delta\phi)^2}$. An additional requirement on the track isolation transverse momentum (p_T^{iso}) which is the sum of the p_T of tracks from the primary vertex in a cone of $\Delta R = 0.2$. The pileup dependence of the track isolation transverse momentum is reduced by selecting tracks consistent with the originating diphoton production vertex. Two selected photons passing the object definitions are required to satisfy $E_T^{\gamma_1}/m_{\gamma\gamma} > 0.4$ and $E_T^{\gamma_2}/m_{\gamma\gamma} > 0.3$. The isolation transverse energy requirement is $E_T^{iso} < 0.22 \times E_T^{\gamma} + 2.45$ GeV and track isolation transverse momentum must satisfy $p_T^{iso} < 0.05 \times E_T^{\gamma}$.

6. Statistical analysis

6.1. Introduction

The yields for the signal and background events are obtained by performing an unbinned maximum likelihood fit on the diphoton invariant mass distribution for each mass hypothesis for both the LWA and NWA. The entire mass spectrum is used for every mass hypothesis tested. The asymptotic approximation [10] is used to calculate the background-only local p-value (p_0). A global significance is computed with the look-elsewhere-effect (LEE) taken into account based on the number of positive crossing of the 2σ significance levels in the region $m_X \in [200, 2000]$ GeV. The expected and observed 95% CL exclusion limits are computed using the asymptotic approximation described in Ref [10].

6.2. Background model

The background is modelled using a data driven approach to estimate the continuous background contribution. It is modelled using a smooth function of the form of an exponential. The simplest functional form containing the least degrees of freedom is used and has the form:

$$f_0(x; b, a_0) = (1 - x^{1/3})^b x^{a_0} \qquad (2)$$

where a_0 and b are allowed to be free floating and $x = \frac{m_{\gamma\gamma}}{\sqrt{s}}$.

6.3. Signal model

The signal is parametrised using a DSCB (See Equation 1) that has been characterised using fully simulated ggF samples for both NWA and LWA for various mass hypotheses. Each parameter of the DSCB is fitted with a polynomial as a function of the mass hypothesis allowing a continuous description of the signal shape for all mass hypothesis values.

6.4. Systematics

Systematic uncertainties are treated as nuisance parameters that are introduced into the likelihood functions in the background-only and signal-plus-background fits. There are a large number of systematic sources such as object uncertainties (photon energy resolution), theoretical sources (production cross sections) and fitting errors (Signal shape and background shape).

Source	Uncertainty
	Background modeling $^{\circ\bullet}$
Spurious signal	$2-10^{-3}$ events, mass-dependent
Background fit	$\leq 50\%-\leq 20\%$ of the total signal yield uncertainty, mass- and signal-dependent
	Signal modeling $^{\circ\bullet}$
Photon energy resolution	$^{+[55-110]\%}_{-[20-40]\%}$, mass-dependent
	Signal yield $^{\bullet}$
Luminosity	$\pm 5\%$
Trigger	$\pm 0.63\%$
	C_X *factors* $^{\bullet}$
Photon identification	$\pm(3-2)\%$, mass-dependent
Photon isolation	$\pm(4.1-1)\%$, mass-dependent
Production process	$\pm 3.1\%$

Table 1: Summary of the systematic uncertainties in the signal-plus-background likelihood fit when considering the NWA signal model. The $\circ$ symbol indicates categories of uncertainties that affect the local p-value for the background-only hypothesis, while the $\bullet$ symbol denotes uncertainties that impact the limit on $\sigma_{\text{fiducial}} \times \text{BR}(X \to \gamma\gamma)$.

Table 1 shows a summary of the uncertainty contributions when considering the NWA signal model.

7. Results

The background only likelihood fit to the observed data from ATLAS is shown in Figure 1. Figure 2 a) shows the ATLAS results for the local p-value for the background-only fit for the NWA and Figure 2 b) shows the corresponding 95% CL for the NWA. The largest deviation is observed at about 750 GeV with a local significance of 3.6σ which is reduced to 2.0σ when the LEE affect is taken into account. CMS performed a similar study in Run I and II [11] and measured a combined p-value which has the largest deviation at 750 GeV with a significance of 3.0 σ which is reduced to 1.7 σ after taking into account the LEE.

8. Discussion and Conclusions

A search for scalar resonances in the diphoton channel was performed by ATLAS with the 2015 dataset consisting of 3.2 fb^{-1} of pp collisions. The largest deviation from the background-only hypothesis occurs in the 750 GeV region with a global significance around 2σ. The limits exceed those set in Run I. The excess at 750 GeV was also observed by the CMS Collaboration Final clarification on this excess is expected to happen with more data taken during the 2016 physics period.

References

[1] The ATLAS Collaboration 2012 *Physics Letters B* **716** 1–29 URL `http://www.sciencedirect.com/science/article/pii/S037026931200857X`
[2] The CMS Collaboration 2012 *Physics Letters, Section B: Nuclear, Elementary Particle and High-Energy Physics* **716** 30–61
[3] Veltman M J G and Yndurain F J 1989 *NPB* **325** 1

IOP Conf. Series: Journal of Physics: Conf. Series **802** (2017) 012011 doi:10.1088/1742-6596/802/1/012011

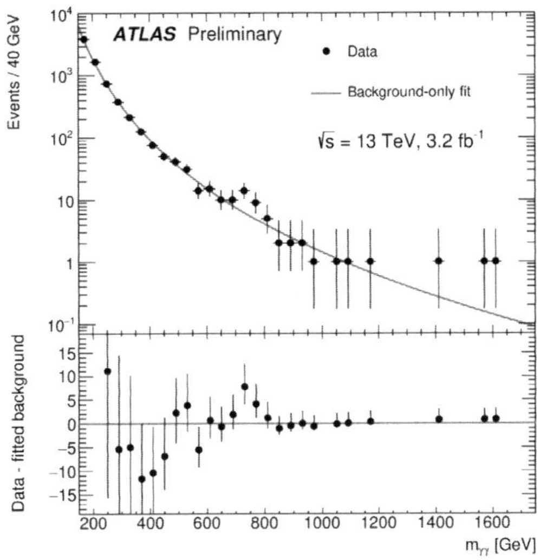

Figure 1: The invariant mass distribution for ATLAS with the background-only fit overlaid.

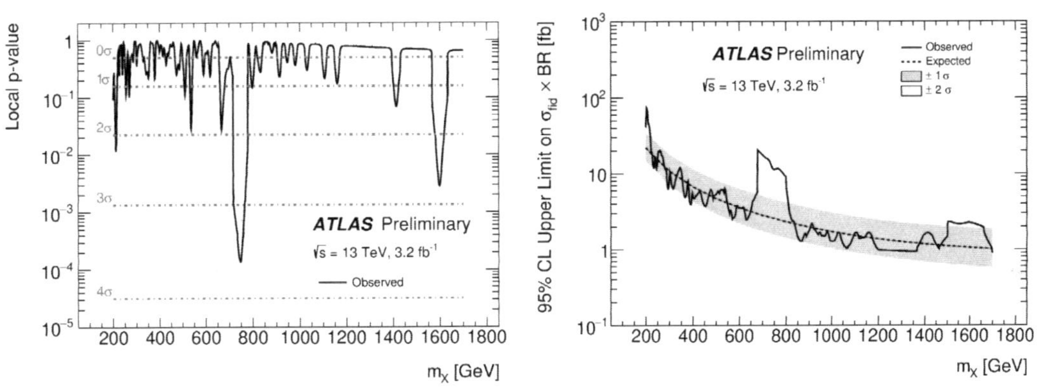

(a) Background p-value

(b) 95 % CL limits

Figure 2: ATLAS results for a) p-value for the background-only hypothesis p_0 as a function of the mass m_X of a probed NWA resonance signal and b) Expected and observed upper limits on $\sigma_{\text{fiducial}} \times \text{BR}(X \rightarrow \gamma\gamma)$expressed at 95% CL, as a function of the assumed value of the narrow-width scalar resonance mass.

[4] Binoth T and van der Bij J J 1997 *ZfPC* **75** 17–25
[5] Schabinger R and Wells J D 2005 *PRD* **72** 093007
[6] ATLAS Collaboration 2015 Search for resonances decaying to photon pairs in 3.2 fb^{-1} of pp collisions at $\sqrt{s}$ = 13 TeV with the ATLAS detector Tech. Rep. ATLAS-COM-CONF-2015-096 CERN Geneva URL `https://cds.cern.ch/record/2111338`
[7] ATLAS Collaboration 2008 *JINST* **3** S08003
[8] CMS Collaboration (CMS) 2015 *PLB* **750** 494–519
[9] ATLAS Collaboration 2014 *Eur. Phys. J.* **C74** 3071
[10] Cowan G, Cranmer K, Gross E and Vitells O 2011 *EPJC* **71** 1554
[11] 2015 Search for new physics in high mass diphoton events in proton-proton collisions at $\sqrt{s}$ = 13 TeV Tech. Rep. CMS-PAS-EXO-15-004 CERN Geneva URL `https://cds.cern.ch/record/2114808`

IOP Conf. Series: Journal of Physics: Conf. Series **802** (2017) 012012 doi:10.1088/1742-6596/802/1/012012

Performance of Jet Vertex Tagger in suppression of pileup jets and E_T^{miss} in ATLAS detector

K G Tomiwa[1*]

[1] School of Physics, University of the Witwatersrand, Johannesburg 2050, South Africa

E-mail: kehinde.gbenga.tomiwa@cern.ch

Abstract. The Intermediate missing transverse analysis group are investigating the production of Higgs with missing transverse energy. This analysis is based on missing transverse momentum (E_T^{miss}) in ATLAS detector and Jets are important objects in reconstruction of missing transverse momentum in ATLAS experiment. In run I, the Jet E_T^{miss} group used the jet vertex faction method to discriminate pileup jet from hard-scatter. Due to increase in collusion energy in the ATLAS run II a new method called the Jet vertex tagger has been developed. This document presents the performance of Jet Vertex Tagger (JVT) working points using ATLAS simulated Monte Carlo samples. Comparing the fraction and efficiency of jets passing JVT cuts or run I and run II, we see about 20% difference in run II and run I.

1. Introduction

The Large Hadron Collider (LHC) is an hadronic collider used to collide proton beams. The LHC was designed for search for physics in support of the Standard Model (SM) and new physics for the theories Beyond the Standard Model (BSM). Collisions of proton bunches in LHC result in hard-scatter (HS) interaction know as signal and an addition collision called the pile-up (PU) collision. The presence of these simultaneous interactions (PU) dilute the signal in LHC. Simulations show that the number of PU interactions will increase in the LHC Run 2 to about ~ 40 as compared to ~ 25 in Run 1 and more after upgrade to High Luminosity LHC [1]. The LHC experiment is handling this challenge by: improving the basic event reconstructions, removal of pile-up before event final state and improving the physics interpretation tools like b-tagging and jet-suppression techniques to be more effective against pile-up. This document presents the performance of jet using the improved jets vertex tagger algorithm.

2. ATLAS Detector

The LHC has different detectors at different positions along the ring. These detectors and the experiments associated with them are called ATLAS (A Toroidal LHC Apparatus), CMS (Compact Muon Solenoid), ALICE (A Large Ion Collider Experiment), LHCb (Large Hadron Collider Beauty), TOTEM, LHCf and MoEDAL. Each experiments has unique setup and sizes. The ATLAS detector is a general purpose detector that is used to investigate different range of physics. The detector has sub-system like the inner detector ($|\eta| < 2.5$)used mainly for vertex reconstruction, electromagnetic and hadronic sampling calorimeters used to reconstruct energy in the detector and muon system. The ATLAS detector covers pseudorapidity range of $|\eta| < 4.9$.

IOP Conf. Series: Journal of Physics: Conf. Series **802** (2017) 012012 doi:10.1088/1742-6596/802/1/012012

The detector has special trigger system used for event selection. A detailed description can be found in Ref. [2]

3. Motivation

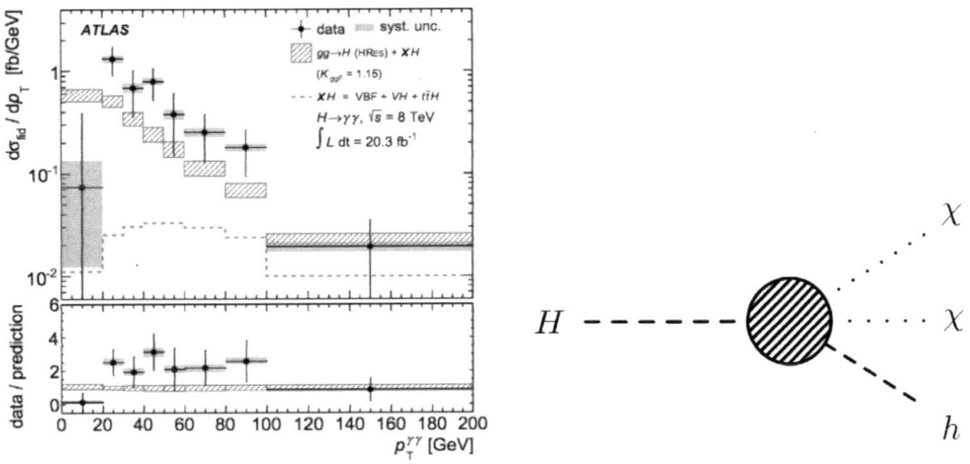

Figure 1. Left: Higgs p_T spectrum Run 1 [3]. Right: Decay of a Heavy scalar particle [4]

Figure 1 Left shows the transverse momentum (p_T) data for Higgs production at center of mass energy of 8 TeV, the shaded regions depict the theoretical prediction and the dots represent the experimental data. The plots shows a difference in the p_T spectrum of observation and SM prediction.

Among all the possible explanations of Higgs boson p_T behavior, is the production of Higgs in associated with invisible particles such as Dark Matter (DM) candidate. Figure 1 Right shows one of the predicted decay mode of an Heavy scalar particle decaying to the SM Higgs and DM candidates. The DM candidate are inferred by the presence of E_T^{miss} in the detector. The dominant decay modes are $b\bar{b}$, WW*, ZZ*,$\tau\tau$, $\gamma\gamma$ in decreasing order of branching ratio. None of these channels predicted the Higgs boson p_T distribution like the one observed in ATLAS, this could be because of E_T^{miss} from particle not observable by the detector.

4. E_T^{miss} and E_T^{miss} Reconstruction

The missing transverse energy (E_T^{miss}) is an imbalance in the sum of energy of all particles traveling transverse to the beam axis. This value is the negative vector sum of the momenta of all particle detected in the proton-proton collisions. E_T^{miss} reconstruction is performed by ATLAS using the energy deposited in the calorimeter and the Muon spectrometer [5]. The E_T^{miss} components are calculated by:

$$E_{x(y)}^{miss} = E_{x(y)}^{miss,calo} + E_{x(y)}^{miss,\mu} \tag{1}$$

The calorimeter term in Equation 1 is calculated based on reconstructed physics objects (electrons, photons, taus, jets and muons) associated with cells, cells which are not associated with physics objects are tagged $E_{x(y)}^{miss,CellOut}$. The $E_{x(y)}^{miss,calo}$ term is given by:

$$E_{x(y)}^{miss,calo} = E_{x(y)}^{miss,calo,e} + E_{x(y)}^{miss,calo,\gamma} + E_{x(y)}^{miss,calo,\tau} + E_{x(y)}^{miss,calo,jets} + E_{x(y)}^{miss,calo,\mu} + E_{x(y)}^{miss,CellOut} \quad (2)$$

each term in Equation 2 is calculated from the negative sum of the calibrated cell energies corresponding to objects as follows:

$$E_x^{miss,term} = -\sum_{i=1}^{N_{cell}} E_i sin\theta_i cos\phi_i, \qquad E_y^{miss,term} = -\sum_{i=1}^{N_{cell}} E_i sin\theta_i cos\phi_i \quad (3)$$

The muon term in Equation 1 is calculated from the momenta of muon reconstructed with $|\eta| < 2.7$ as follows:

$$E_{x(y)}^{miss,\mu} = -\sum_{muons} P_{x(y)}^{\mu} \quad (4)$$

E_T^{miss} and its angle ϕ^{miss} are calculated by:

$$E_T^{miss} = \sqrt{(E_x^{miss})^2 + (E_y^{miss})^2}, \qquad \phi^{miss} = arctan(E_y^{miss}, E_x^{miss}) \quad (5)$$

A good E_T^{miss} performance is gotten when using E_T^{miss} calibration with the following configuration:

- $E_T^{miss,e}$ term is calculated using medium electron with $p_T > 10$ GeV [6].
- $E_T^{miss,\gamma}$ term is calculated from "tight" photon with $P_T >$ 10GeV.
- $E_T^{miss,\tau}$ term is calculated from "tight" taus [7] with $P_T >$10 GeV and with the LCW calibration
- $E_T^{miss,soft}$ term is calculated from jets with $7 < |P_T| < 20$ GeV (reconstructed using the anti-kt algorithm with R=0.6) calibrated with the LCW calibration
- $E_T^{miss,jets}$term is calculated from jets with $P_T > 20$ GeV (reconstructed using the anti-kt algorithm with R=0.6) calibrated with the LCW calibration and the jet energy scale factor (JES) [8].

The $E_T^{miss,soft}$ is the energy sum of all energy clusters or tracks that are not associated with any of the identified physics objects, this makes the $E_T^{miss,soft}$ most sensitive to pile-up which is corrected by implementation of JVT.

5. Jet tagger

The ATLAS experiment uses Jet-Vertex-Tagger (JVT) variables to identify and suppress PU jets coming. The JVT is a replacement for the old discriminate: jet vertex fraction (JVF) Equ. 6 variable used Run I. The JVT variable is a 2-dimensional likelihood variable, it is estimated based on the k-nearest-neighbour algorithm whose inputs are corrected-JVF (`corrJVF`) Equ. 7 and R_{pT} Equation 8 [1].

$$JVF = \frac{\sum_k P_T^{trk_k}(PV_0)}{\sum_k P_T^{trk_l}(PV_0) + \sum_{n\geq 1}\sum_l P_T^{trk_k}(PV_n)}. \quad (6)$$

$$corrJVF = \frac{\sum_k P_T^{trk_k}(PV_0)}{\sum_k P_T^{trk_l}(PV_0) + \frac{\sum_{n\geq 1}\sum_l P_T^{trk_k}(PV_n)}{(k.n_{trk}^{PU})}}. \quad (7)$$

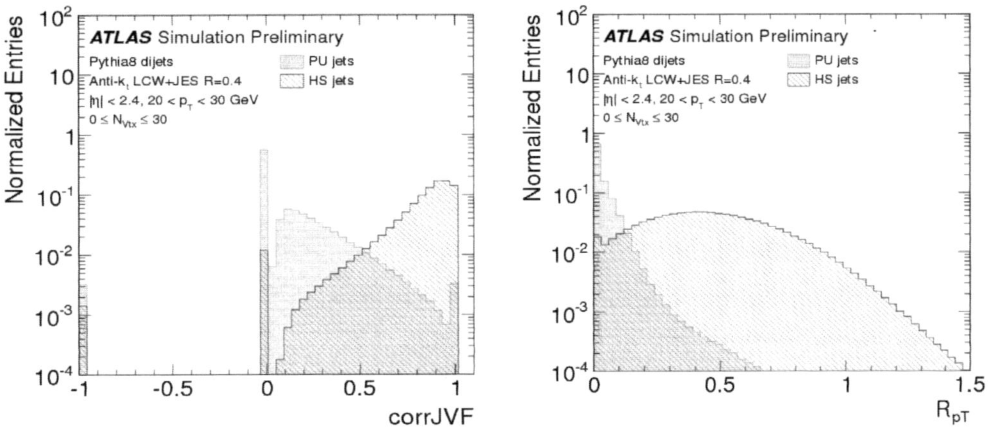

Figure 2. Left: corrJVF distribution for pileup and hard-scatter jet in simulated dijet events. corrJVF = -1 for jets with no associated tracks. Right: R_{pT} distributions for PU and HS jets from Ref. [1]

$$R_{pT} = \frac{\sum_k p_T^{trk_k}(PV_0)}{P_T^{jet}}. \tag{8}$$

From Figure 2, showing the distribution of coorJVF and R_{pT} in simulated sample, we see HS jets are characterized by large corrJVF and large R_{pT} and PU jets are concentrated at low R_{pT} and low corrJVF values. The jet E_T^{miss} recommendation defines three JVT working point (JVT WP), which are JVT cut 0.14, JVT cut 0.64 (default) and JVT cut 0.92

6. Performance of the new jet tagger

Using the fraction (the proportion of all hard-scatter divided by the pileup jet in all jet passing a given JVT cut) and efficiency (the proportion of hard-scatter/pileup jet passing given JVT cut) distributions to study the performance of JVT.

6.1. Fraction

The Hard-scatter and Pileup jet fraction is defined as:

$$F_{HS} = \frac{N_{HS}}{N_{HS} + N_{PU}}, \qquad F_{PU} = \frac{N_{PU}}{N_{HS} + N_{PU}}, \tag{9}$$

where $F_{HS/PU}$ represents the fraction of hard-scatter/pileup jet, $N_{HS/PU}$ is the number of hard-scattered or pileup jet.

6.2. JVT Performance

Figures 3 and 4 show fraction jet distribution for gluon-gluon fusion sample at center of mass energy of 13 TeV comparing with Run 1 JVF 8 TeV sample. Figure 3 shows the fraction of jet before events are corrected for number of primary vertex. The Run 2 jet distributions without JVT cut differ from the Run1 by about 20% in both PU and HS and the default JVT WP ranges between 50 - 65% in the HS fraction, 10 - 50% for PU fraction. Figure 4 shows the fraction of

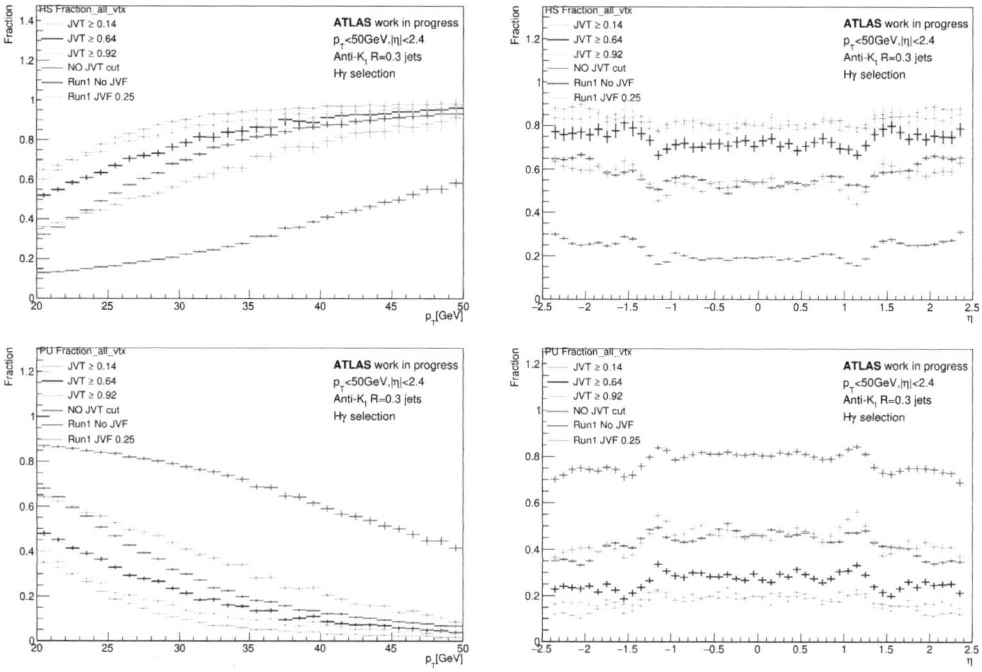

Figure 3. Distributions of hard-scatter (top row) and pileup (bottom row) jet fraction in JVT region as a function of (left) p_T and (right) η from the gluon-gluon fusion Higgs production mode when all vertices are considered.

jet after events are corrected for number of primary vertex by rejecting event outside $z < 0.3mm$ of the detector. The Run 2 distribution without JVT cut improves as compared with the Run1 by about 10% in both PU and HS, There is good agreement between jet performance for JVT 0.64 (Run II) and JVF 0.25 JVF (Run I). Purity of jets passing the JVT 0.64 (default JVT WP) cut ranges from 60% to 85% VS P_T in the HS and 40% to 15% VS P_T in the PU.

7. Conclusions

A search for an Heavy scalar particle which decays to Higgs with associated Dark Matter particle which are inferred by the presence of intermediate E_T^{miss} in the detector.

The search is heavily influenced by E_T^{miss}, and E_T^{miss} is in turn sensitive to the jet going into its reconstruction algorithm. A new jet-vertex tagging method, JVT, has been developed to mitigate the effect of pile-up jets. The JVT algorithm is the improved version of the previous JVF algorithm used in ATLAS. The JVT algorithm is based on two new pileup-insensitive variables: corrJVF and R_{pT} in 2-dimension likelihood.

the selection of events with correct vertex improves the distribution of fraction of jets for all the JVT working point. Purity of jets passing the JVT 0.64 cut ranges from 60% to 85% VS P_T. There is good agreement between jet performance for JVT 0.64 (Run II) and JVF 0.25 JVF (Run I). Significant amount of Pileup jets in the detector's forward region $(0.5 \leq |\eta| \geq 1.5)$ were observed as shown by the purity vs η plots.

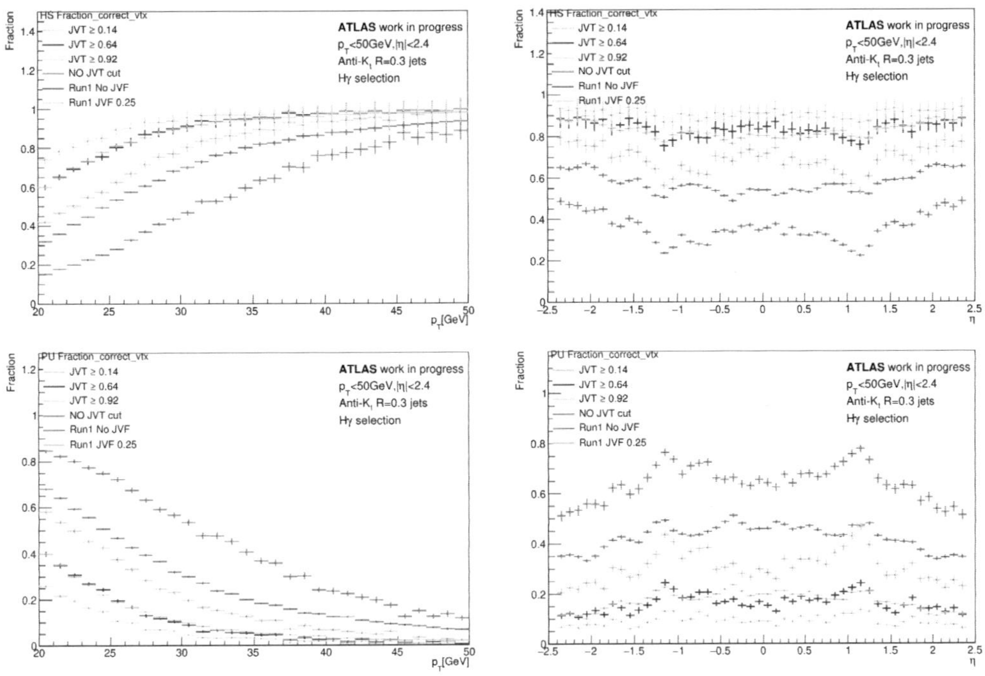

Figure 4. Distributions of hard-scatter (top row) and pileup (bottom row) jet fraction in JVT region as a function of (left) p_T and (right) η from the gluon-gluon fusion Higgs production mode when all vertices are considered.

References

[1] Aad G *et al.* 2014 Tagging and suppression of pileup jets with the ATLAS detector Tech. Rep. ATLAS-CONF-2014-018 CERN Geneva URL `https://cds.cern.ch/record/1700870`

[2] Aad G *et al.* 2008 *Journal of Instrumentation* **3** S08003 URL `http://stacks.iop.org/1748-0221/3/i=08/a=S08003`

[3] Aad G *et al.* (ATLAS) 2014 *JHEP* **09** 112 (*Preprint* `1407.4222`)

[4] von Buddenbrock S 2015 *Journal of Physics: Conference Series* **645** 012017 URL `http://stacks.iop.org/1742-6596/645/i=1/a=012017`

[5] Aad G *et al.* (ATLAS Collaboration) 2011 Reconstruction and Calibration of Missing Transverse Energy and Performance in Z and W events in ATLAS Proton-Proton Collisions at 7 TeV Tech. Rep. ATLAS-CONF-2011-080 CERN Geneva URL `https://cds.cern.ch/record/1355703`

[6] Aad G *et al.* (ATLAS Collaboration) 2010 Data-Quality Requirements and Event Cleaning for Jets and Missing Transverse Energy Reconstruction with the ATLAS Detector in Proton-Proton Collisions at a Center-of-Mass Energy of $sqrts = 7$ TeV Tech. Rep. ATLAS-CONF-2010-038 CERN Geneva URL `https://cds.cern.ch/record/1277678`

[7] Aad G *et al.* (ATLAS Collaboration) 2010 Tau Reconstruction and Identication Performance in ATLAS Tech. Rep. ATLAS-CONF-2010-086 CERN Geneva URL `http://cds.cern.ch/record/1298857`

[8] Aad G *et al.* (ATLAS) 2013 *Eur. Phys. J.* **C73** 2304 (*Preprint* `1112.6426`)

IOP Conf. Series: Journal of Physics: Conf. Series **802** (2017) 012013 doi:10.1088/1742-6596/802/1/012013

Neutral meson and direct photon analysis with ALICE

A.M. Whitehead for the ALICE Collaboration
Department of Physics, University of Cape Town, Rondebosch 7701, South Africa
E-mail: andile.mothegi.whitehead@cern.ch

Abstract. The measurement of neutral mesons, particularly π^0's and η's, plays an important role in the study of the Quark-Gluon Plasma (QGP), the hot and dense medium created in high-energy heavy-ion collisions. Parton energy loss in the QGP, often called jet quenching, can be assessed via measuring the suppression of high-$p_{\rm T}$ π^0's in heavy-ion collisions, when compared to pp collisions using the nuclear modification factor (R_{AA}). Furthermore, neutral mesons are the dominant source of photons in pp and Pb-Pb collisions, and their precise measurement is required to measure direct photons that are produced thermally within the QGP or in hard initial scatterings in the earliest phases of the collision. In both cases, high-quality measurements in pp collisions are required as a reference for Pb-Pb collisions. ALICE measurements of neutral meson spectra cover a large $p_{\rm T}$ range, with the Photon Conversion Method - which requires measurements from the ITS and TPC - covering low to intermediate $p_{\rm T}$ and the PHOS and EMCal electromagnetic calorimeters covering an intermediate to high $p_{\rm T}$ range. In this presentation, measurements of π^0's and η's obtained from the ALICE experiment, for pp collisions at several collisional center of mass energies ($\sqrt{s_{\rm NN}}$), from 0.9 TeV to 8 TeV and in Pb-Pb collisions at $\sqrt{s_{\rm NN}} = 2.76$ TeV, will be presented. The reconstruction of neutral mesons using the Photon Conversion Method (PCM) will also be discussed.

1. Introduction

High energy heavy-ion collision experiments are believed to produce an extremely hot, dense phase of matter comprised of deconfined partons, known as the Quark-Gluon Plasma (QGP) [1, 2]. The ALICE Experiment [3] at CERN's Large Hadron Collider (LHC) [4] is mainly concerned with the study of heavy-ion collisions, particularly, the study of the QGP.

The verification of the QGP and characterisation of its properties is a complex procedure requiring the analysis of several complementary measurements. Among the observables used to probe this evolving state of matter is the measurement of direct photons. These are photons that are emitted from the evolving system as opposed to those photons from hadronic decays. Since photons only interact electromagnetically with the strongly coupled medium, and considering the size of said medium, the information they bear pertaining to the properties of the system at the time of their production remains relatively undistorted [5]. There are numerous mechanisms responsible for the production of direct photons. Direct photons of particular interest are prompt photons from initial hard scattering processes [6] and those from thermal scatterings [7] in both the QGP and hot hadron gas phases of the medium. In A-A collisions these different processes are expected to have varying contributions to direct photon production depending on the analysed p_T regions. Prompt photon production - from quark-gluon scattering and quark-anti-quark annihilation - calculable via next-to-leading order (NLO) perturbative Quantum Chromodynamics (pQCD)[8], is expected to dominate at high p_T. On the other hand, thermal photons are expected to dominate direct photon production in A-A collisions at low p_T ($p_T \lesssim 4$ GeV/c). The inclusive photon signal is dominated by a large background attributable to hadron decays, of which ~80% and ~18% is due to neutral pion and η mesons, respectively [9].

Hadron production in pp and A-A collisions at $p_T \gtrsim 3$ GeVc, is attributable to initial hard scatterings which subsequently undergo jet fragmentation [10]. High energy partons traversing a QGP will experience energy loss due to parton-medium interactions. These energy losses are manifested as suppressed hadron production in the high p_T region. A measure of this suppression is given by the particle nuclear modification factor,

$$R_{AA} = \frac{d^2N/dp_T dy|_{AA}}{\langle T_{AA}\rangle \times d^2\sigma/dp_T dy|_{pp}},$$

where the nuclear overlap function, $\langle T_{AA}\rangle = \langle N_{coll}\rangle / \sigma_{pp}^{inel}$ is related to the average number of inelastic nucleon-nucleon collisions [11] with $\langle N_{coll}\rangle$ being the number of nucleon-nucleon binary collisions and σ_{pp}^{inel} being the pp inelastic cross-section. Measurement of the R_{AA} can give insight into the mechanisms of parton energy loss and hence allow for an improved characterisation of the QGP.

2. Measurements

The ALICE detector is well equipped for the measurement of photons. The ALICE Photon Spectrometer (PHOS) [13] is a highly segmented electromagnetic calorimeter comprised of $PbWO_4$ crystals, with a coverage in pseudorapidity and azimuth of $|\eta| < 0.13$ $\Delta\varphi = 60°$, respectively. PHOS measures electromagnetic showers created by photons interacting with the scintillator material. The purity of photon samples measured by PHOS is ensured by performing cuts on the energy, size and dispersion of the particle electromagnetic shower. Another means of measuring photons is by using the Photon Conversion Method (PCM); this reconstructs converted photons within the detector via tracking the e^+e^- conversion products. e^+e^- pairs from conversion photons originate from secondary vertices some distance from the primary vertex and are said to have a V^0 topology - due to the charge neutrality of the conversion photon and V-shaped topology of the secondary track pairs. The low conversion probability of photons (~8.5%) in the detector is compensated by the large acceptance ($|\eta| < 0.9$ and $\Delta\varphi = 2\pi$ coverage) of the Inner Tracking System (ITS) [3] and the Time Projection Chamber (TPC) [14] which together track the e^+e^- pairs. High purity photon samples are achieved by performing cuts on the decay topology and on charged track properties. Details of these measurements for pp and Pb-Pb at $\sqrt{s_{NN}} = 2.76$ TeV are presented in [15].

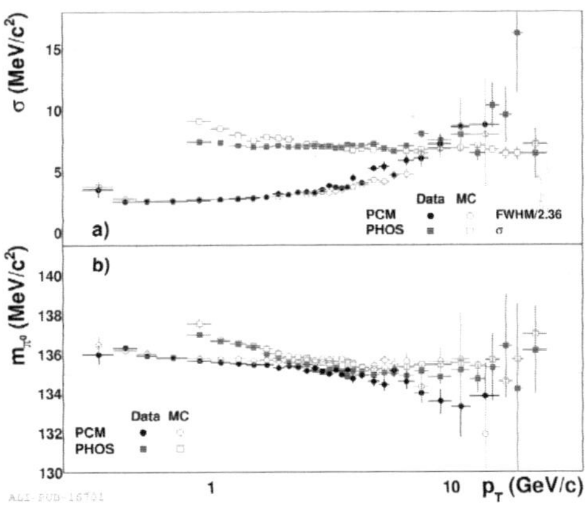

Figure 1. Reconstructed π^0 peak width (a) and position (b) versus p_T in pp collisions at $\sqrt{s} = 7$ TeV for both PHOS and PCM. Measurements are compared to Monte-Carlo simulations. The horizontal line in (b) represents nominal π^0 mass [12].

The measurement of neutral mesons, particularly π^0s, is an important measurement in its own right [12]. Extracting the direct photon signal requires the subtraction of the decay photon spectrum from the inclusive photon spectrum. π^0 and η decays, which constitute the bulk of the decay photon spectrum, are reconstructed by calculating the invariant mass of photon pairs using,

$$M_{\gamma\gamma} = \sqrt{2E_{\gamma 1}E_{\gamma 2}\left(1 - \cos\theta_{1,2}\right)},$$

where $E_{\gamma 1}$ and $E_{\gamma 2}$ represent the two photon energies, and $\theta_{1,2}$ describes the opening angle between them.

As indicated in Fig. 1, PHOS is particularly adept at reconstructing the π^0 in the mid to high p_{T} range. PCM, on the other hand, exhibits an exceptional energy resolution from low to mid p_{T} with a π^0 peak width of $\sigma_{\pi^0} \approx 3$ MeV/c at low p_{T}. The combination of PCM and PHOS enables the reconstruction of π^0 and η mesons over a large p_{T} range.

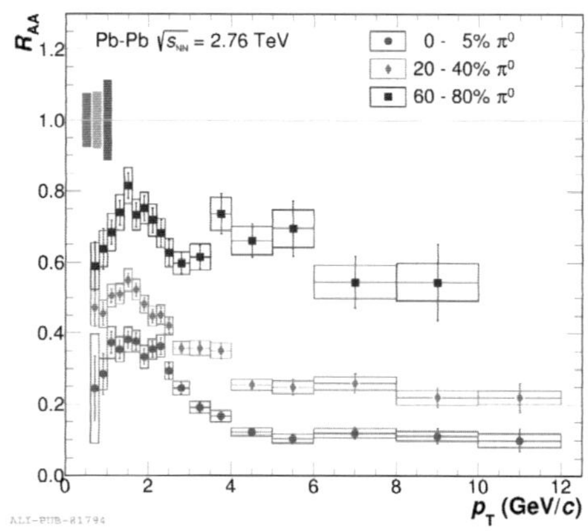

Figure 2. π^0 R_{AA} for Pb-Pb collisions at $\sqrt{s_{\mathrm{NN}}} = 2.76$ TeV for centrality classes 0-5%, 20-40% and 60-80%. Vertical bars and boxes represent statistical and systematic uncertainties respectively. Horizontal bars describe the bin width. The boxes around unity describe the uncertainty of the average nuclear overlap function $\langle T_{AA}\rangle$ and the normalization uncertainty of the pp spectrum added in quadrature [16].

Much of the difficulty in measuring the direct photon signal is in obtaining a good estimate of the decay photon signal. The decay photon spectrum is calculated using a Monte Carlo based approach, to which the parametrised π^0 - and if available the η - spectrum is used as input. The remaining hadron contribution to the decay spectrum is estimated using m_T scaling [17]. Merely subtracting the decay photon spectrum from the inclusive photon spectrum results in large uncertainties in the direct photon spectrum. These uncertainties are primarily attributable to certain systematic uncertainties in the measurement procedure. In [15] the main sources of systematic uncertainties for PCM are attributable to the material budget of the detector ($\sim 4.5\%$) and the photon selection criteria which is as high as $\sim 4\%$ at low p_{T}. The main systematic uncertainties in the case of PHOS are the uncertainties in the energy scale of the detector ($\sim 9.5\%$) and on the efficiency of the reconstruction calculation, which is as high as $\sim 3\%$ in the $0 - 20\%$ centrality class. To minimise these uncertainties, one calculates the direct photon spectrum in the following manner:

$$\gamma_{\mathrm{dir}} = \gamma_{\mathrm{inc}} - \gamma_{\mathrm{decay}} = \gamma_{\mathrm{inc}}\left(1 - \frac{\gamma_{\mathrm{decay}}}{\gamma_{\mathrm{inc}}}\right), \tag{1}$$

where the reciprocal of the above fraction is redefined as a double ratio, given by:

$$R_\gamma \equiv \frac{\gamma_{\mathrm{inc}}}{\pi^0_{\mathrm{param}}} / \frac{\gamma_{\mathrm{decay}}}{\pi^0_{\mathrm{param}}}, \tag{2}$$

IOP Conf. Series: Journal of Physics: Conf. Series **802** (2017) 012013 doi:10.1088/1742-6596/802/1/012013

where π^0_{param} is the parametrised π^0 spectrum. The systematic uncertainties characteristic of each fraction in Eq. (2) partially cancel [15]. A R_γ value greater than unity indicates a direct photon signal. The double ratio in Eq. (2) is used in Eq. (1) to calculate the direct photon spectrum.

3. Results

Fig. 2 shows the combined π^0 R_{AA} for PHOS and PCM for three different centrality classes. In all centrality classes the R_{AA} exhibits a peak in the region $1 \lesssim p_{\text{T}} \lesssim 2$ GeV/c, and decreases in the $2 \lesssim p_{\text{T}} \lesssim 3 - 6$ GeV/c region followed by a flattening out of the R_{AA} at larger p_{T}. In the high p_{T} region, where π^0 production is expected to be dominated by fragmentation of hard scattered partons, an increase in suppression with increasing centrality is observed. In the high p_{T} region, an R_{AA} of $\sim 0.5 - 0.7$ is observed in the 60-80% centrality class, and is as low as ~ 0.1 in the 0-5% centrality class. Given the expectation that more central collisions create a larger medium, one would expect that at high p_{T} where parton energy loss due to hard parton-medium interactions should be dominant, a larger medium being traversed would result in an increase in parton energy loss and hence an increase in suppression at high p_{T}. These measurements are in agreement with these expectations.

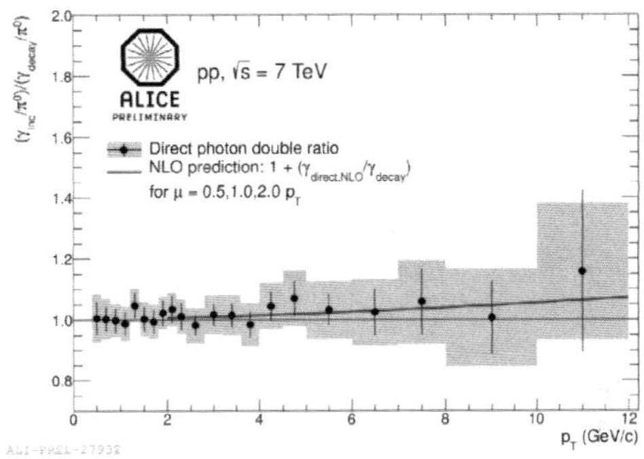

Figure 3. The measured double ratio in pp collisions at $\sqrt{s} = 7$ TeV compared to NLO pQCD predictions [9].

In Fig. 3, a measurement of an excess above the decay photon signal, represented by the double ratio, R_γ, for pp collisions at $\sqrt{s} = 7$ TeV along with NLO pQCD calculations of the double ratio are shown. This measurement provides a baseline for the double ratio measurement performed in Fig. 4 [9]. Although measurements and NLO pQCD calculations are in agreement, the large systematic uncertainties, do not allow for a decisive conclusion on the existence of direct photon production in pp collisions at $\sqrt{s} = 7$ TeV.

The left side of Fig. 4 shows the double ratio for three centrality classes for Pb-Pb collisions at $\sqrt{s_{\text{NN}}} = 2.76$ TeV. All centrality classes exhibit a direct photon excess in the $p_{\text{T}} \gtrsim 4$ GeV/c range. The 0-10% centrality class also exhibits an excess in the $1 \lesssim p_{\text{T}} \lesssim 4$ GeV/c range. The double ratio in all centralities is compared to three NLO pQCD calculations [15] - which predict the prompt photon contribution to the direct photon yield - scaled up according to the number of expected binary collisions in each class [18]. In all centrality classes, pQCD calculations are in agreement with the measurements in the $p_{\text{T}} \gtrsim 5$ GeV/c region. The excess of ~10-15% for 0-20% centralities - not predicted by the scaled pQCD calculations - in the $0.9 \lesssim p_{\text{T}} \lesssim 2.1$ GeV/c suggests the presence of another source of direct photons in central collisions. The excess for each data point in this p_{T} region is $\sim 2\sigma$ above unity. The combined significance of the excess in the $0.9 \lesssim p_{\text{T}} \lesssim 2.1$ GeV/c region is found to be 2.6σ [15]. The right hand side of Fig. 4 shows the direct photon spectra for the three centralities, calculated from the double ratios on the left. The direct photon spectra are compared to four state of the art direct photon calculations, over and beyond the entire p_{T} range of the data. All models assume the formation of a QGP and include the contribution of pQCD photons. The primary difference between the models is their treatment of the space-time evolution of the system. The van Hees et al [19] calculation is

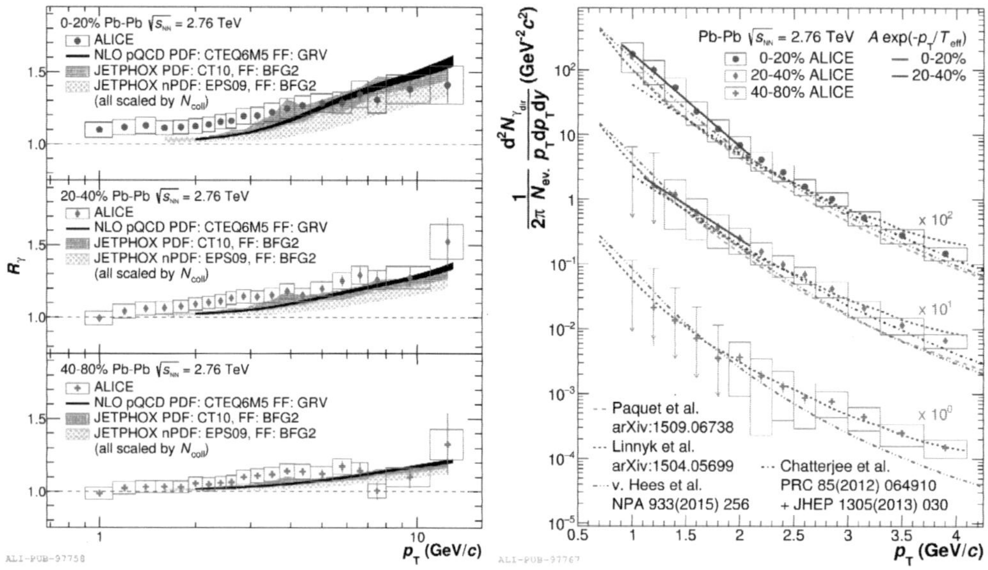

Figure 4. (left) The measured double ratio in Pb-Pb collisions at $\sqrt{s_{\mathrm{NN}}} = 2.76$ TeV for 0-20%, 20-40% and 40-80% centrality classes. The measured double ratio spectra are compared to scaled NLO pQCD predictions as well as JETPHOX event generator predictions. (right) Comparison of several modelss with direct photon spectra in Pb-Pb collisions at $\sqrt{s_{\mathrm{NN}}} = 2.76$ TeV for the 0-20% (scaled by a factor 100), the 20-40% (scaled by a factor 10) and 40-80% centrality classes. Contribution of pQCD photons is included in all models. In addition, the more central 0-20% and 20-40% classes are fitted with an exponential function [15].

based on ideal hydrodynamics. The Chatterjee et al [20] model is based on a (2+1D) event-by-event ideal hydrodynamics calculation with fluctuating initial conditions under the assumption of longitudinal boost invariance. The Paquet et al [21] calculation is similar to the Chatterjee calculation; however, instead of ideal hydrodynamics, viscous hydrodynamics is used. The Linnyk et al. [22] calculation is based on an off-shell transport approach in which the full evolution of the collision is described microscopically. The models also adopt different initial conditions of the system; namely, the time at which the space-time evolution of the system begins and the intitial temperature at the centre of the fireball. Unfortunately, a comparison of the effectiveness of the models cannot be determined as all models fall within the systematic uncertainties of the measurements. Consequently, a reduction of systematic errors is required. Although they agree within the uncertainties, at low p_{T} for the most central class, the models appear to slightly underestimate the data.

For a static medium in thermal equilibrium the distribution of the thermal radiation is described by Planck's law, which in the case of sufficiently high energy photons can be approximated by Wien's approximation described by an exponential. Consequently, models in which thermal photon production is the dominant source of direct photons at low p_{T}, make the argument that an exponential function $\propto \exp(-p_{\mathrm{T}}/T_{\mathrm{eff}})$ fitted to this region of the direct photon spectrum is a valid procedure [7], where the inverse slope parameter T_{eff} can be interpreted as a characterisation of the effective temperature of the medium averaged over its entire space-time evolution [23]. However, because the medium is cooling and expanding rapidly at speeds of up to $2/3c$, one would have to integrate over the states of the system during its space-time evolution and take into account the the red- and blue-shifting of the emitted photons. This in general should not result in an exponential distribution. As such, the relationship between T_{eff} and the mean temperature of the system is less direct. This complicates the interpretation of T_{eff}, but a correlation between the slope and the initial temperature still exists [24]. In the two most central classes, over the region $0.9 < p_{\text{T}} < 2.1$ GeV/c, the pQCD calculation by Paquet et al. was subtracted from the direct photon measurement, and to this region an exponential of the form $A\exp\left(-p_{\text{T}}/T_{\text{eff}}\right)$ was fitted. The extracted inverse slope parameter in the most central class, $T_{\text{eff}} = 297 \pm 12^{\text{stat}} \pm 41^{\text{syst}}$ MeV significantly exceeds the requisite critical temperature $T_{\text{c}} = 155 - 170$ MeV for creating a QGP [25, 26].

4. Conclusion

Neutral pion and direct photon measurement with ALICE using PHOS in conjunction with PCM, allows for a large coverage in p_{T}. In addition, at high $p_{\text{T}} \gtrsim 6$ GeV/c, a large suppression in π^0 production in most central Pb-Pb collisions at $\sqrt{s_{\text{NN}}} = 2.76$ TeV is consistent with parton energy loss. For $p_{\text{T}} \gtrsim 5$ GeV/c, measurements of the direct photon yield, for all centralities, are in agreement with NLO pQCD calculation, scaled by the number of binary nucleon-nucleon collisions. Although only achieving a 2.6σ significance, an excess in direct photon production at $p_{\text{T}} \lesssim 4$ GeV/c, for the most central classes was observed. In 0-10% centralities, an exponential fit to the $0.9 < p_{\text{T}} < 2.1$ GeV/c region was found to have an inverse slope parameter $T_{\text{eff}} = 297 \pm 12^{\text{stat}} \pm 41^{\text{syst}}$ MeV. Consequently, the measurement of thermal photons in heavy-ion collisions at larger collision energies will be performed with ALICE Pb-Pb data at $\sqrt{s_{\text{NN}}} = 5.02$ TeV. Greater statistics should reduce statistical uncertainties, however; a better handle on systematic uncertainties is necessary to achieve more precise measurements.

References

[1] S. Borsnyi *et al.*, Physics Letters B **730**, 99 (2014).
[2] HotQCD Collaboration, A. Bazavov *et al.*, Phys. Rev. D **85**, 054503 (2012).
[3] The ALICE Collaboration, K. Aamodt *et al.*, Journal of Instrumentation **3**, S08002 (2008).
[4] L. Evans and P. Bryant, Journal of Instrumentation **3**, S08001 (2008).
[5] M. H. Thoma, Phys. Rev. D **51**, 862 (1995).
[6] T. Ferbel and W. R. Molzon, Rev. Mod. Phys. **56**, 181 (1984).
[7] S. Turbide, R. Rapp, and C. Gale, Phys. Rev. C **69**, 014903 (2004).
[8] D. J. Gross and F. Wilczek, Physical Review D **9**, 980 (1974).
[9] M. Wilde, Nuclear Physics A **904905**, 573c (2013), The Quark Matter 2012Proceedings of the {XXIII} International Conference on Ultrarelativistic NucleusNucleus Collisions.
[10] A. Majumder and M. van Leeuwen, Progress in Particle and Nuclear Physics **66**, 41 (2011).
[11] K. Eskola, R. Vogt, and X.-N. Wang, International Journal of Modern Physics A **10**, 3087 (1995).
[12] B. Abelev *et al.*, Physics Letters B **717**, 162 (2012).
[13] G. Dellacasa *et al.*, CERN/LHCC , 99 (1999).
[14] J. Alme *et al.*, Nuclear Instruments and Methods in Physics Research Section A: Accelerators, Spectrometers, Detectors and Associated Equipment **622**, 316 (2010).
[15] J. Adam *et al.*, Physics Letters B **754**, 235 (2016).
[16] B. Abelev *et al.*, The European Physical Journal C **74**, 1 (2014).
[17] P. K. Khandai, P. Shukla, and V. Singh, Phys. Rev. C **84**, 054904 (2011).
[18] K. Aamodt *et al.*, Physical review letters **105**, 252301 (2010).
[19] H. van Hees, M. He, and R. Rapp, Nuclear Physics A **933**, 256 (2015).
[20] R. Chatterjee, H. Holopainen, T. Renk, and K. J. Eskola, Phys. Rev. C **85**, 064910 (2012).
[21] J.-F. Paquet *et al.*, arXiv preprint arXiv:1509.06738 (2015).
[22] O. Linnyk, V. Konchakovski, T. Steinert, W. Cassing, and E. Bratkovskaya, Physical Review C **92**, 054914 (2015).
[23] C. Shen, U. Heinz, J.-F. m. c. Paquet, and C. Gale, Phys. Rev. C **89**, 044910 (2014).
[24] D. dEnterria and D. Peressounko, The European Physical Journal C-Particles and Fields **46**, 451 (2006).
[25] Y. Aoki, Z. Fodor, S. Katz, and K. Szab, Physics Letters B **643**, 46 (2006).
[26] HotQCD Collaboration, A. Bazavov *et al.*, Phys. Rev. D **85**, 054503 (2012).

Institute of Physics
Dirac House, Temple Back
Bristol BS1 6BE UK

ISSN: 1742-6588
ISBN 978-1-5108-3694-5

CURRAN ASSOCIATES INC.
proceedings.com